MW01486973

THE GREAT CACTI

THE SOUTHWEST CENTER SERIES

JOSEPH C. WILDER, SERIES EDITOR

THE GREAT CACTI

Ethnobotany & Biogeography

David Yetman

THE UNIVERSITY OF ARIZONA PRESS
TUCSON

Maps drawn by Paul Mirocha

The University of Arizona Press
© 2007 The Arizona Board of Regents
All rights reserved

Library of Congress Cataloging-in-Publication Data

Yetman, David, 1941–
   The great cacti : biogeography and ethnobotany of
columnar cacti / David Yetman.
      p. cm. — (The southwest center series)
   Includes bibliographical references and index.
   ISBN 978-0-8165-2431-0 (hardcover : alk. paper)
   1. Saguaro—Southwestern States. 2. Cactus—South-
western States. 3. Cactus—Mexico. 4. Cactus—South
America. 5. Ethnobotany. 6. Cactus—Utilization.
I. Title.
QK495.C11Y477 2007
583'.56—dc22   2007013777

Publication of this book is made possible in part by a grant from
the Southwest Center of the University of Arizona.

The Southwest Center Series list is on page 299.

Manufactured in Korea on acid-free, archival-quality paper.

12 11 10 09 08 07 6 5 4 3 2 1

To Linda Wallace-Gray, a true lover of cacti.

To all the Guarijíos, Mayos, and Seris who taught me of the importance of columnar cacti. They remain the world's true experts.

CONTENTS

ACKNOWLEDGMENTS

Without the ongoing financial support of Agnese Haury, I could not have begun this book, so great were the travel demands. My department director at the University of Arizona's Southwest Center, Joe Wilder, tolerated my frequent absences from the office while I made forays into various deserts to photograph. My wife, Lynn Fowler, joined me in several trips and consented to be photographed with the great cacti. Vicente Tajia, who shares my passion for cacti, educated me (or attempted to) in the lore of the *pitaya* and *etcho* cacti. Enriquena Bustamante generously shared with me the data from her detailed and disciplined studies on the growth and reproduction of *Stenocereus thurberi*. She also helped me locate the grove of *Pachycereus grandis* near Coyula, Oaxaca.

My colleagues from the Public Broadcasting System program *The Desert Speaks*—Dan Duncan, Tom Kleespie, and Yar Petryszyn—were patient and understanding as they waited while I photographed. Dan helped me locate plants in Brazil. Roberto Neumann provided me with information on *Trichocereus atacamensis* and *Cereus validus*. Bill Risner shared my enthusiasm in the Huastecan lowlands of eastern Mexico. Dr. Carlos Ostolaza introduced me to some of his favorite cactus haunts in Peru, as did Fátima Cácares. Jens Madsen guided me to several important species in Ecuador. Dan Austin read parts of the manuscript and suggested additional references as well as important changes to the writing. Jeff Banister, who appears frequently in the photographs in this book, accompanied me through much of southern Mexico. Graham Charles made important comments on my identification of Peruvian cacti. He and Dr. Ivor Crook made traveling in Oaxaca an even greater pleasure. Martin Lowry identified some Bolivian species from my photographs. Arnoldo Michel Rosales helped me find excellent *Neobuxbaumia squamulosa* plants.

My niece Faline Harshbarger and my nephew Benton Yetman also traveled to far-flung places with me, as did my brother Richard and my friends and colleagues Alberto Búrquez and Fritz Jandrey. Axel Nielsen assisted me in Argentina and Bolivia. Helia Bravo-Hollis, whom I never met, died at the age of one hundred while I was writing this book. Her pioneering publications on Mexican cacti forged the way for much of my traveling.

THE GREAT CACTI

1

The
GREAT CACTI

The offspring of the ancient marriage of plants and people
are far stranger and more marvelous than we realize.

—MICHAEL POLLAN,
THE BOTANY OF DESIRE

Figure 1.1. Saguaros in flower, Tucson Mountains, Arizona.

OVER THE PAST FIFTY YEARS, I have become a willing victim of *cactophily*—the inordinate love of cacti. As a desert rat in the cactus-rich Sonoran Desert of Arizona, I frequently run across a fair cross section of native cacti—prickly pears, barrels, jumping chollas, hedgehogs, claret cups, rainbows, and pincushions. They compose a curious lot, some of them with magnificent flowers, all of them quite willing to puncture one's skin. My interest in these lesser plants wanes, however, in comparison with my fascination for the huge saguaros with their eerie likeness of the human form. (Fig. 1.1.)

Until my first trip to Mexico in the early 1960s, my familiarity with great cacti was limited to saguaros, organ pipes, and the odd *sinita (Pachycereus schottii)* of Organ Pipe Cactus National Monument. South of the border I caught glimpses from my motor scooter of a bewildering variety of new and strange species—giants, unfamiliar races as large as the saguaro, but decidedly different. Over the years, I learned of other varieties, all blessed with their own individual presence, their signature. As I studied native cultures of my home state and Mexico, I discovered that wherever the great cacti grew, indigenous people found them not only noticeable, but useful as well.

Beginning in the late 1960s, I had the good fortune to spend several years off and on living among the Seris of the Gulf of California coast of northern Sonora, Mexico. Decades later I spent even more time a couple of hundred miles farther south, studying the lands and plants of the Mayos of northwest Mexico. Shortly thereafter I ventured into the adjacent mountainous lands of the Guarijíos of southern Sonora and came to know them. All of these folk have made their homes in the region for centuries. I found that for all these people of the desert and semidesert, columnar cacti were their most important plants—the *sahueso* or *cardón (Pachycereus pringlei)* for the Seri, the organ pipe or *pitaya (Stenocereus thurberi)* for the Mayo, and the *etcho (Pachycereus pecten-aboriginum)* for the Guarijío.[1] Most of them would agree that without columnar cacti, they could hardly have survived in their lands and become the people they are. After a few decades of walking and talking with people so close to the ground, I was struck with the notion that writing about the giants of the cactus family and of the various ways people used them might be an important step in understanding cacti themselves, getting to know the people among whom they grow, and appreciating the natural environments where they have evolved. It might also further the cause of cactus conservation.

In this book, I describe most of the world's columnar cacti, dwelling on those that native peoples have found to be most beneficial, but focusing as well on those that

Figure 1.2. Pachycereus pringlei *and* boojums *(Fouquieria columnaris), Cataviña, Baja California.*

Figure 1.3. Opuntia echios, *Galápagos Islands, a tall cactus, but not a columnar.*

are the most spectacular. I provide especially detailed descriptions of cacti found in northwestern Mexico and the southwestern United States, for they live and effuse their influence within a day's drive of my home in Tucson, Arizona. For my purposes, a columnar cactus is simply one that is rather tall and ribbed and is usually taller than it is wide, a plant in which verticality clearly predominates over horizontality and in which the trunks and branches (arms) are roughly cylindrical. This characterization requires subjective judgments at times, and other cactologists use different definitions, but it will do for the purposes of this book. Research by Robert Wallace has established that columnar cacti share certain molecular characteristics, so grouping them according to their shape is not arbitrary.[2] Mexican cactologist Alejandro Casas has pointed out that Nahua (Aztec) speakers in central and southern Mexico used the term *nochcuauitl*—literally translated "tree cactus"—to set off columnars from other cacti. We have no such term. Cactologists sometimes refer to them as *cereoids,* a name derived from early cactus taxonomy when all columnar cacti were placed in the genus *Cereus.* (Figs. 1.2, 1.3, and 1.4.)

Figure 1.4. Barrel cactus (Ferocactus diguetii), *Santa Catalina Island, Gulf of California. Although some plants of the species grow more than 2 m tall, too many are shorter and disqualify it from columnarhood.*

Figure 1.5a. Escontria chiotilla *near Reyes Metzontla, Valle de Tehuacán, Puebla.*

Figure 1.5b. Trunk of Pachycereus weberi, *Las Flores, Puebla.*

My disproportionate attention to the columnars of northwest Mexico is the result of a fluke of biogeography and human history: nowhere else do so many cacti constitute such culturally important plants. South America has several great cacti of pivotal cultural importance, as do central and southern Mexico and a handful of Caribbean islands. But the peculiar conditions of human migrations, aridity, heat, and cactus evolution in the states of Arizona and Sonora have concentrated more ethnobotanical cactus tales there than in any other area.

The great cacti are extraordinary plants, strange and, to many, alluring and mysterious. Without leaves but usually armed with potent spines that replace them, they reach about 20 m (66 ft.) in height and weigh as much as 25 tons. Their sensuous flowers, thick with stamens, may exceed 15 cm (6 in.) in diameter. One plant may host hundreds such flowers and produce more than 25 kg (60 lbs.) of fruit each year.[3] At least two species (*Neobuxbaumia mezcalaensis* and *N. polylopha*) grow more than 15 m (50 ft.) tall as a single, solitary stalk, and other species reach a similar height and resemble a candelabrum of a hundred arms or branches. Other kinds branch at the base into a proliferation of columns or segments. Some are nearly spine free, whereas others are shielded with tens of thousands of potent spines more than 15 cm long. Some have trunks 3 m tall and more than a meter in diameter, whereas others are trunkless, branching from the base, with no branch more than 20 cm thick. (Figs. 1.5a and 1.5b.)

Popular stories of toweringly tall columnar cacti resemble reports of enormously long anacondas and pythons. I have read of specimens of *Carnegiea gigantea, Cereus jamacaru, C. lamprospermus, Pachycereus grandis,* and *P. pringlei* that exceeded 20 m (66 ft.) in height. Mexican researchers inform me that they have measured a 20-m-high *Pachycereus weberi* in Oaxaca, perhaps the one illustrated in figure 1.8, which may be the tallest *P. weberi* yet recorded. A published book contains a photo of a *P. grandis* purportedly 24 m tall—that's 80 ft.[4] A photograph taken in the year 2000 shows a sahueso from Baja California nearly 19 m (63 ft.) tall.[5] A *fallen* saguaro was measured at 20.72 m (68 ft.) tall.[6] The plant explorer Howard Scott Gentry reported a *sahuira (Stenocereus montanus)* in southern Sonora that he measured at 20 m using a "*pitahaya* pole" as a measuring stick.[7] The tallest cacti I have tried to measure are a *clavija (Neobuxbaumia mezcalaensis)* in Oaxaca, Mexico, apparently in excess of 20 m tall and the aforementioned *chico (Pachycereus weberi)* from Oaxaca. A *tetecho (Neobuxbaumia tetetzo)* near Calipan, Puebla, may be nearly as tall. I have examined closely a fine *Neobuxbaumia polylopha* in Mexico's Barranca de Metztitlán that *may* have been well more than 16 m (54 ft.) tall. Several sahuesos I have seen in Baja California are nearly as tall, and natives assure me taller ones are found in more remote canyons. The tallest plants probably grow near canyon bottoms, where they receive protection from strong winds and lightning. (Figs. 1.6, 1.7, 1.8, 1.9, and 1.10.)

Figure 1.6. Unusually tall Neobuxbaumia tetetzo, *near Calipan, Valle de Tehuacán, Puebla.*

Figure 1.8. Pachycereus weberi *roughly 18 m (60 ft.) tall, Quiotepec, Oaxaca.*

Figure 1.7. View from base to crest of 20-m-tall (66 ft.) Neobuxbaumia mezcalaensis *near Huajuapan de León, Oaxaca.*

Figure 1.9. Pachycereus weberi *near Calipan, Valle de Tehuacán, Puebla.*

Figure 1.10. Cephalocereus totolapensis *in excess of 13 m (43 ft.) tall, Totolapan, Oaxaca.*

In addition to being sensational oddities, the great cacti are plants of immense importance to New World cultures, providing food, shelter, medicine, and experientially delicious and religiously significant hallucinogens. It is as if apple, peach, and pear trees grew wild, were a source of protein, and, with no tending, yielded fruit, medicine, lumber, industrial products, and sublime psychoactive liquids. Brazilians not only eat the fruits of the *mandacarú (Cereus jamacaru)*, but consider it a reliable weather prognosticator as well.

Perhaps most important to humans is the ability of giant cacti to flourish in areas that are marginal or unsuitable for productive agriculture—the slopes too steep; the climate too dry, hot, or cold; soils too poor, rocky, or salty. Noteworthy are the huge *pasacanas (Trichocereus atacamensis)* in arid northwestern Argentina that flourish in rocky, thin, barely developed soils on very dry slopes and in valleys at high altitudes where freezing is common. One would hardly expect to find giant plants of tropical origin in such punishing habitats, but they are numerous and widespread. Legions of these great sentinels stand out in this harsh terrain, as if they employed alchemy to transform dry, cold matter from bitingly frigid air and forbidding landscapes

into food and building materials. The same can be said for other columnars, such as the *soberbios (Browningia candelaris)* of rainless northern Chile and southern Peru that achieve great size where no other plant appears to grow, and sahuesos of the very dry portions of the Sonoran Desert in Baja California and Sonora. And so it is with the great cacti everywhere. They are transubstantiators on a grand scale. (Fig. 1.11.)

I saw my first large columnar cactus at night. My family had moved from the eastern U.S. coast when I was twelve to a tiny town in eastern Arizona located not in the Sonoran Desert, but in the periphery of the Chihuahuan Desert, which has no columnar cacti. On one momentous trip, my father and I arrived in Tucson from the east at dusk and drove to visit friends to the north of the city. From my earliest memories, I had longed to see real saguaros, which to a New Jersey hayseed lad were as exotic as tropical fig trees. I had marveled about them from photographs and descriptions in *National Geographic* magazines. I scanned both sides of the road as dusk fell, but the highway leading into Tucson from the east had none of the great cacti growing nearby. We drove through the city and took the narrow highway leading to the hilly north side, perfect

Figure 1.11. Stunted Trichocereus atacamensis *at 3900 m (13,000 ft.), Salar de Uyuni, Bolivia.*

saguaro habitat. As we left the valley and climbed into the foothills, I caught my first glimpse of saguaros as a series of eerie silhouettes on the western horizon. Then headlights caught the monster plants on the roadside, and the shadows played with the odd angles and colors of the cacti, accentuating their stature and weirdness. I could hardly speak from excitement. From that time on, I wanted to be—to live—where the big cacti grow. They were my special plants. It was to be seven years before I could do so. To this day I live among columnar cacti, eighteen species of them in pots or planted in the ground in my yard. One of them provides me with dozens of sweet, tangy fruits each summer. Two more will also supply me when they reach maturity in a few decades. That prospect makes old age worth waiting for. (Fig 1.12.)

In the fifty-some years since my first glimpse of a saguaro, I have wandered extensively among columnars. As abundant as they are, they have never lost their magic for me. I have helped myself to their fruits and seeds, even used their flesh to heal. On many occasions, I have sought refuge from the searing summer sun in their dense shade, which is at least equal to that of most conventional leafy trees. The massive canopy of *Pachycereus weberi*—the

tenchanochtli cardón, its Aztec name, also called *chico*, a name applied with characteristic native irony—would afford more protection in a rainstorm than almost any broadleaved tree. But my enchantment with the giant cacti is also aesthetic. I find myself enthralled by the singing sounds when the wind whips around and over their spines and by the swaying of the tall ones in the wind, almost like palm trees. I still wonder that they never seem to topple over in spite of their apparent top heaviness. (They do, though, and soon thereafter the fallen giants rot and become small ecosystems in the desert.) To this day I consider the most magical outdoor night of my life one I spent in 1990 camping alone in the rich—but in this case open—thornscrub of southern Sonora. My campsite lay on a vast, gentle *bajada* (alluvial fan) among a stand of tall etchos and pitayas interspersed with low trees of the dry forest. A full moon accentuated the legions of spines that cover the etcho fruits, decorating the great plants like silver and golden Christmas ornaments. The *monte*, padded with a soft carpet of native grasses, was alive with discrete rustlings of elusive creatures of the night. Owls glided silently by overhead, hunting the source of these intriguing noises.

Figure 1.12. Flowering saguaros, Sonoran Desert National Monument, Arizona.

That place is now much changed. Exotic buffelgrass (*Pennisetum ciliare*), introduced from Africa, has invaded the open spaces, threatening the ancient forest with fires, and has choked out native grasses and shrubs, thus eliminating the domestic resources needed by a host of wild animals. Tough, coarse, closely spaced clumps of buffelgrass have replaced the flat carpet of low grasses that made for ideal camping. The cacti survive, but I fear they will soon be burned or felled to make way for more open pasture where more cows can be fattened. In Baja California, hundreds of hectares thick with sahuesos are bulldozed each year to make way for fields irrigated with water pumped from a source that will be depleted in a couple of decades. In Ecuador, giant cardones (*Armatocereus cartwrightianus*) are felled, burned, and chopped up into cattle feed during the droughts. In east central Brazil, clearing for pasture and for soybeans raised for livestock feed is even more rapid, and the habitats of the myriad hoary *Pilosocereus* and other small columnars are vanishing. The giants must be sacrificed before the golden calf. We who love the great cacti are on a collision course with a hamburger-hungry world.

Columnar cacti are widespread—growing from the Sonoran Desert in Arizona, Baja California, and Sonora through the thornscrub and tropical deciduous forests of northwestern Mexico; from the Chaco forests to the Andean deserts of Argentina and into the rainless Atacama of Chile; through the frigid altiplano of Peru and Bolivia to the semiarid valleys of Ecuador, the cloud forests of the upper Amazon, and the thorny vastness of Brazil's northeast *sertão,* all the way to the Caribbean. I have trudged through cactus woodlands of Mexico, Peru, and the Galápagos. In every case, these profound plants have increased my wonder at their immensity and variety and at their usefulness to humanity. (Fig. 1.13.)

I am not alone in my admiration of great cacti. The saguaro flower is the state flower of Arizona. Visitors and natives alike flock to the desert to gape at these odd, intriguing trees. Especially dense cactus populations have been protected in national parks in Arizona and Argentina. The United Nations has designated the incomparable Valle de Tehuacán and adjacent Valle de Cuicatlán as Biosphere Reserves, and Mexico has made an especially dense grove of columnar cacti into a national botanical garden. Chile has set aside two national parks where columnar cacti grow on fog alone. Thousands of Brazilians raise mandacarús in their yards. Some of the world's finest cactus gardens are in Europe, where no cactus existed until the conquering

Figure 1.13.
Browningia
pilleifera, *Balsas,*
Río Marañon,
northern Peru.
The green patch
to the right of the
river is avocado
and mango
orchards irrigated
with river water.

Europeans brought back seeds and plants from the New World. (Fig. 1.14.)

Many columnar cacti resemble the human figure. Their branches are commonly called "arms" both in English and in Spanish. Great ranks of hillside *viejitos* of the Valle of Tehuacán of Puebla, Mexico, so resemble legions of soldiers that botanists assigned them the scientific name *Cephalocereus columna-trajani,* after the military formations of the Roman emperor Trajan. Locally they are called "old men," from their stooped aspect, and a *Pilosocereus* with cottony tufts at the top is called *viejita* (old woman). Stands of pasacanas resemble multitudes crowding into the canyons of the Andes. The human resemblance may also bring out our darker side: in 1982 north of Phoenix, a drunken off-road traveler tried to fell a saguaro by ramming it with his Jeep. The stubborn cactus refused to fall. In a rage, the vandal pulled a shotgun from his gun racks and emptied both barrels into the resistant but hapless plant. The drunk succeeded. The blast weakened the base of the cactus. It toppled over on top of him, crushing him to death. David Grundman was his name, and he came to be celebrated in derisive ballads. The mills of the cactus gods grind slowly. (Fig. 1.15.)

Uses of the Great Cacti

Xote das Meninas

Mandacarú quando fulorá na seca
E um siná que a chuva chega na sertão
Toda menina que enjoa da boneça
E sinal de que o amor já chegou no coração . . .

—GILBERTO GIL

The Mandacarú when it flowers in the drought
It is a sign that rain is coming to the sertão
For every girl who wearies of her doll
It is a sign that love has come to her heart . . .

Brazilians are often incurable romantics, seeing in the flowers of the mandacarú *(Cereus jamacaru)* each October and November a sure sign of the imminence of rain to temper the withering heat of the sertão, along with blossoming love. A less poetic observer, a prominent botanist, has written, "Apart from their wide appeal to specialist growers and collectors of the unusual, cacti have few uses."[8] Not only do the author's words exemplify eurobotanico-

Figure 1.14. The saguaro flower is the state flower of Arizona.

Figure 1.16. Welcome monument with Cereus jamacaru, Bom Jesus de Lapa, Bahia, Brazil.

Figure 1.15. Cephalocereus columna-trajani, Texcala, Valle de Tehuacán, Puebla.

centrism (European plant prejudice), but he is dead wrong. Cacti have as many uses as most other plant families, or more—for food, medicine, lumber, fencing, medicine, toys, trinkets, and hallucinations. Cactus fruits and flesh should be celebrated as a food source as much as corn, potatoes, and tomatoes are. For many species, horticulture requires only that the planter cut off a short piece, allow it to dry for a few days, and then stick it in the ground. No grafting, backcrossing, or detasseling is required, as well as no

weeding, no cutting canes, no spraying with chemicals, and no lacing the soil with growth hormones. Over the years, the amateur or professional horticulturalist may select the most productive plants for propagation, but apart from that mild human intervention, farmers need add almost nothing. (Fig. 1.16.)

Native Americans can testify to the significance of columnar cacti. The most important native nondomesticated plant for the Tohono O'odham (formerly called the Papago)

Figure 1.17. Crested Cereus repandus, *Aruba.*   *Figure 1.18. Cactus fence of* Stenocereus griseus, *Bonaire.*

of Arizona and Sonora is the saguaro. For the Seris of northwest Mexico, it is the sahueso; for the Mayos and Yaquis, the organ pipe or pitaya *(Stenocereus thurberi);* for Guarijíos, the etcho *(Pachycereus pecten-aboriginum);* and for Cáhita-speaking peoples of the Río Fuerte in Sinaloa, Mexico, the sahuira *(Stenocereus montanus).* The same can probably be said of the pasacana for the pre-Incan people of northwest Argentina and southwest Bolivia, the tetecho for aboriginal peoples of the Valle de Tehuacán of Puebla in southern Mexico, and a variety of pitayos *(Stenocereus* spp.) for the native peoples of Oaxaca. For native islanders of Aruba, Bonaire, and Curaçao, it was the *kadushi (Cereus repandus).* (Fig. 1.17.)

The buds and flowers of many columnar cacti may be cooked and eaten, and the branches of one species may be made into soup or stew. The fruits are usually sweet, nutritious, and abundant. Some, such as *saleas* (tetecho fruits), are a tasty dietary staple. Sweet or not, the fruits provide important nutrition and dietary enjoyment for thousands of people. Their seeds yield oil and flour and constitute basic nutrition for indigenous groups. The fruits of great cacti have saved lives by alleviating the ravages of scurvy.[9] Amputated and planted, their limbs grow into living fences. From their flesh come medicine, dye, beverage bases, shampoo, and a spiritual catalyst. Alive, they offer cooling shade in the desert. Dead, they surrender their wood as lumber of remarkable versatility and as firewood. For those seeking spiritual elevation, they open the doors of otherworldly perception. (Fig. 1.18.)

In the human landscape, great cacti offer shelter from sun and wind. Individual plants form natural landmarks, a sort of cultural anchor. Without wilting or browning, they endure climates that produce scorching heat, blasts of frost, and withering drought. They require no watering, no fertilizers, no chemical additives, no genetic manipulation, yet they are amenable to cultivation. Other than old age and its vicissitudes, their only natural enemies are lightning strikes, humans, and the animals, machines, pollution, and plagues that humans loose upon them.

It is not only humans that have benefited far back in their history from the offerings of columnars. The cacti are a vital source of food to myriad animals. Many birds, insects, and mammals rely nearly exclusively on cactus flowers and fruits for at least part of the year. One moth species *(Upiga virescens)* cannot live outside the sinita (senita), and the cactus itself cannot pollinate without the moth.[10] Coyotes and White-winged Doves are two larger vertebrates for whom the fruits of pitayas and saguaros are indispensable. Coyotes overcome the spiny exteriors and ingest huge quantities of the pulp. Around my hometown of Tucson in late June and early July, desert roadways are littered with reddish coyote scats consisting almost entirely of undigested saguaro and later prickly pear fruit seeds. Foxes and raccoons are also avid devourers of the food. In Sonora, the same is true for pitaya and sahueso fruits. If the cactus topples when it dies, the enormous carcasses often rot, ferment, and liquefy, becoming host to a virtual ecosystem of bacteria, fungi, arthropods, reptiles, and small

Figure 1.20. *Pitayas, fruits of the pitaya (organ pipe),* Stenocereus thurberi, *Carbó, Sonora.*

Figure 1.19. Neobuxbaumia tetetzo, *near Zinacatepec, Valle de Tehuacán, Puebla.*

mammals. Some bats have developed a symbiotic relationship with cactus flowers. Irreplaceable pollinators, many migrating long-nosed bats, *Leptonycteris curasoae,* require the nectar of cactus flowers for survival.

People have often planted the great cacti for their own use. In the widely separated valleys of southern Mexico, natives cultivate and tend at least nine species in addition to the ubiquitous domesticated prickly pears.[11] An additional five species (perhaps more) are deliberately propagated elsewhere in the country. Large cacti function in yards as shade trees and ornamentals, but with no need for pruning, fertilizing, or irrigating. In Argentina, Bolivia, and Chile, archaeological sites can be detected by the presence of large numbers of *Trichocereus atacamensis,* testimony to the importance of the cacti to people long vanished (or to the importance to the cacti of microhabitats created by people!). The Peruvian apple cactus *Cereus hildmannianus* is so popular for its large, spineless fruits that it has spread over most of tropical and neotropical South America, belying its probable wild Argentine, Brazilian, Paraguayan, or Uruguayan origin. It is now a common horticultural addition to many yards in warmer and drier parts of the United States, including mine. The hallucinogenic and virtually spine-free San Pedro cactus (*Echinopsis pachanoi*) is common in pots, hedges, patios, and gardens throughout the Andes and increasingly in yards and pots in the southwestern United States.

Residents of Mexico's Valle de Tehuacán were harvesting cactus fruits and seeds long before they were harvesting corn, at least eight thousand years before the present.[12]

Whether the plants with desirable traits were selected and cultivated prior to the domestication of any grain is difficult to establish, but it is certainly possible. They were not domesticated in the quantity necessary for the development of large-scale civilizations, however. Why corn and not cacti should have become the basic food for North America is difficult to say. At any rate, farming as a fixed occupation seems not to have happened in Mexico until long after corn had been domesticated. Unlike corn and wheat, cacti did not give rise to civilizations. (Fig. 1.19.)

So the great trees have for centuries been providing food for people. They are never troublesome weeds or invaders of property (as are some smaller cacti, such as some prickly pears and chollas [*Cylindropuntia* spp.]). Their spines are quite capable of inflicting painful injuries that are sometimes prone to cause infection, but when one has familiarity with the plants and more practice in handling them, punctures are as avoidable as the toxic chemicals found in unprocessed manioc. Despite their spines, the cacti have worked their way into the larders and pharmacopoeias, ceilings and walls of millions of desert dwellers. (Fig. 1.20.)

Medicinal properties of cacti are important as well. The Badianus Manuscript, written by Aztecs in 1552, recommends a mixture of ashes of the root of *teonochtli* (*Stenocereus* sp., probably *queretaroensis* or *griseus*), deer antler, and certain crushed stones as a cure for dental pain.[13] Modern-day Peruvians consume the large fruits of *sancayos* (*Corryocactus brevistylus*) to cure liver ailments. Application of the flesh of columnars is often recommended for insect bites and stings and for skin disorders.

Figure 1.21. Orchard of Stenocereus pruinosus, *Tehuetzingo, Puebla.*

In one aspect, however, columnar cacti are not useful: few of their stems are edible. Most of them concentrate calcium oxalate in their tissues, a chemical that produces a burning sensation and often nausea if ingested. This characteristic may protect cacti against herbivory, but it is also probably a function of their adaptation to desert soils that are often calcareous. The two genera that are known to carry voluminous quantities of oxalate are both fond of limestone substrates, compounds of calcium. The viejito (*Cephalocereus senilis*) carries this chemical to an extreme: up to 85 percent of its dry weight consists of oxalate.[14]

The exception to the inedibility of columnar cacti that I have documented is the flesh of the kadushi (*Cereus repandus*) of the Netherlands Antilles. To prepare the flesh for eating, the cook removes the green cuticle and separates the outer layer of the flesh from the inner pith. The remaining starchy vegetable is cooked into a stew. However, I suggest that you not generalize from this instance and attempt to prepare meals based on the flesh of columnar cacti. Don't even eat morsels. Save your efforts for the young pads of prickly pears, called *nopalitos,* or wait for the buds, flowers, and fruits of columnar cacti.

With only a few exceptions,[15] these often huge, remarkable plants in their highly varied environments have been largely overlooked or spurned by international plant economists searching for plants with economic potential. Columnar cacti can be significant economic contributors to the food, medicine, and lumber needs of their native lands, the habitats in which they have evolved, and perhaps even elsewhere, as several Mexican scientists have reported in recent decades.[16] The giants are, however, for the most part (except when irrigated) relatively slow growing and equally slow to replace themselves, a disadvantage for rapid food production. (Fig. 1.21.)

Ecology, Columnar Cacti, and Deserts

Each species of columnar cacti—often each variety within a species—has its own demands or requirements, its conditions for living: the right soil, climate, and topography. If these ecological requirements are not fulfilled, a plant may die or fail to flourish, or it may grow, yet fail to reproduce. A saguaro will perish from root rot in the moist tropics; *Pachycereus weberi* will wither away quickly in the long droughts of Baja California. An organ pipe cactus would quickly be killed by freezing in the Andes, but the frost-hardy pasacana of the Andes will burn and die in the blistering heat of the Arizona desert. Saguaros and tetechos often fare poorly in the heavy soils of flatlands, whereas sahuesos and pitayas flourish there. The huge *órganos de cabeza amarilla (Mitrocereus fulviceps)* are endemic (confined) to the precise ecological conditions of a few steep, arid to semiarid canyons in Puebla and Oaxaca, Mexico, at around 1500 m (5000 ft.) elevation. Some special need—whether it be temperature, rainfall, soil, pollinator, or a specific neighbor—is met only in those places and nowhere else. In contrast, etchos seem amenable to growing (sometimes to great size) in a wide variety of frost-free or nearly frost-free habitats with different soils, rainfall conditions, topographies, and plant communities extending from northern to southern Mexico. Argentine pasacanas cannot tolerate rainfall above roughly 250 mm (10 in.) annually and appear to *require* frost and freezing, whereas lower-dwelling but occasionally sympatric (living in the same habitat) cardones *(Trichocereus terscheckii)* welcome rainfall in excess of 700 mm and prefer no more than occasional freezing temperatures. *Unquillos (Stetsonia coryne)* flourish in the seasonally dry Chaco forest of Argentina and Paraguay, where rainfall may reach 1000 mm. The great mandacarú enjoys life in the hot, seasonally dry, and occasionally drought-stricken northeast of Brazil, where *average* rainfall ranges between 500 and 800 mm per year. Frosts may occur in the domain of mandacarús and unquillos. Sinitas cannot tolerate either hard freezes or rainfall greater than roughly 350 mm. *Neobuxbaumia macrocephala* plants are seldom found more than a few to a hectare, whereas their cousins the tetechos crowd more than a thousand strong into the same space.

The majority of columnars is found in deserts or semi-deserts. Popular conceptions of desert habitats envision flowing sand dunes with great cacti (usually resembling saguaros) growing from them. In reality, columnar cacti are seldom if ever found growing naturally in active dunes. I have seen etchos, pitayas, and sahuesos growing from dunes, but only on those that are ancient and stabilized and that form associations with other plants, which in turn help the great plants anchor themselves. On these old dunes, fine particles of airborne silt have mixed with the upper layers of soil to provide a stronger foundation for the sahuesos than do new dunes of pure sand. On the Sonoran coast of the

Figure 1.22. Coastal sand dune encroaching on Pachycereus pringlei, *El Cardonal, Sonora.*

Gulf of California, sahuesos can be seen half buried in the sands of advancing dunes. These old sentinels are doomed by the current dune activity that starves the roots, depriving them of oxygen and moisture and preventing the sun from reaching the green bark and producing photosynthesis. New dunes are gradually overtaking the lower, stabilized dunes, possibly relicts of the Pleistocene, when the climate was cooler and more moist, and they thus threaten a large population of giant sahuesos. (Fig. 1.22.)

At the same location, a few middle-aged sahuesos have suffered the opposite fate: the lee of the dunes has shifted, exposing the medium-size columnars to offshore winds each day from October through June. These relentless winds have stripped formerly cemented sands from the base of the cacti, gradually exposing the root system. Rather than being buried alive, they are being excavated. Propped up only by their increasingly exposed root network, these stalwarts are equally doomed. They inevitably will collapse under their own weight, be blown over in the wind, or die when their roots can no longer deliver needed water and nutrients to them.

Sand dunes do not in themselves make a desert since they are common even in very wet regions. Rainfall and temperature are the critical variables that define deserts. Old-World ecologists often consider deserts only those areas receiving less than 125 mm (5 in.) of rainfall a year. A stringent definition like that would drastically limit our right to call columnar cacti creatures of the desert. Only sahuesos, sinitas, and the *Browningias, Corryocacti, Eulychnias,* and *Weberbauerocerei* of the Chilean and Peruvian Atacama, plus a few pasacanas, pitayas, and saguaros, would qualify as desert cacti. Three eminent cactologists have set the limit at 250 mm (10 in.).[17] By this definition, though, my home in Tucson (300 mm), where saguaros grow nearby, would lie outside the Sonoran Desert. I prefer to use a much looser characterization, considering a desert to be a place where sunshine is abundant; long periods without rainfall are part of the annual cycle; rainfall is greatly exceeded by evaporation; many plant species are adapted to promote water storage, prevent water loss, and withstand drought; and the bare ground is never completely shaded by perennial plants—that is, the vegetation is open, and this open structure is a result of limited rainfall (not of cold temperatures). I know of deserts meeting these conditions that receive more than 375 mm of rainfall.[18] The Antarctic and Artic deserts do not meet them. They have too few plants or none at all. Cold, not scarcity of moisture, limits plant growth at the poles.

Columnars (and all cacti) thrive and proliferate best in horse latitude deserts, regions roughly between 25° and 35° north or south, or in seasonally dry semideserts—that is, places with warm to hot climates and extended periods of drought, such as the sertão in Brazil and the Río Marañon region in Peru. Early in their evolution, cacti abandoned leaves. They carry on photosynthesis through their skin. If they are surrounded by competing plants that block their sunlight, they will grow more slowly or not at all. They also contain large portions of tissues that are specialized for storing water all year round, so hard freezing will damage or destroy these tissues and maim or kill the plants. North of the horse latitudes or at higher elevations, as climates become colder, fewer species of cacti can be found, and plants become smaller and for the most part less common.

Columnar cacti do not appear equally in all deserts of the Western Hemisphere. Their large to massive size, succulent tissues, and need for year-round photosynthesis renders them more or less vulnerable to damage from prolonged cold and absence of sunlight. In the high Andes of northwest Argentina, the columnars *Trichocereus atacamensis, T. tarijensis,* and *Oreocereus* sp. grow well above 3000 m (10,000 ft.) elevation. Frosts and freezing temperatures are common there, but snow and clouds that might keep winter daytime temperatures below freezing are almost unknown (rainfall occurs in the warmer season). Thus, every winter's day brings reliable sunshine, above-freezing temperatures, and the melting of any frost that may have accumulated overnight and potentially caused damage to the plant. The giants are superbly adapted to this environment. They have evolved a flowering pattern that locates flower and fruit on the plants' north sides only, ensuring that the reproductive structures and growing tip will always receive warming sunlight. Such open situations with regular freezing temperatures will kill the huge warmth-loving *Trichocereus terscheckii* of the lower hills and valleys. At its highest elevations, somewhere around 1500 m (5000 ft.), this massive cactus is found only among rocks that retain heat and protect the plants from freezing and on steep slopes where warm air currents rise from the canyon bottoms. Cacti of the genus *Browningia* are found above 1500 m in the very dry (but cool) Atacama Desert of northern Chile and southern Peru, where rains are most unusual, but sunlight is reliable all year round, and frosts and freezes are rare or of short duration. The Chilean columnars of the coastal genus *Eulychnia* survive almost entirely on moisture derived from fog. In some of their range, it has not rained for thousands of years. The landscapes are foggy or hazy much of the year as well, but the fog usually clears off each day and affords the plants full exposure to sunlight. (Fig. 1.23.)

In contrast, in the Chihuahuan Desert[19] of southern New Mexico, western Texas, and a small part of southeastern Arizona, in addition to central and northern Chihuahua, Coahuila, and Durango, Mexico, winter temperatures frequently drop to well below freezing. Except in its extreme southern portion, this vast desert is devoid of columnars, though it is home to more than four hundred species of cacti.[20] Much of it experiences long periods of below-freezing temperatures and may see days pass with no sunshine. Columnars, with their substantial mass and corresponding vulnerability to freezing, would not survive such conditions. Saguaros, the quintessential columnar cactus of the Sonoran Desert, can tolerate frost, even some freezing, with only localized damage, but not the extremes found in the Chihuahuan Desert. At some time during the daylight hours, they must experience above-freezing temperatures, or they begin to die off rapidly.[21] With prolonged freezing temperatures, massive die-offs of saguaros may occur, and many survivors of blasts of cold air bear constriction rings in their trunks where freezing has scarred the outer tissues and left a permanent mark.

The increasing frequency of such deep freezes marks the northern and northeastern limits of the saguaro. I recall as a youth the drive from my hometown of Prescott, Arizona, in Ponderosa pine uplands south to Phoenix in the Sonoran

Figure 1.23. Fog-watered Eulychnia saint-pieana, *Llanos de Challe National Park, Chile.*

Desert. I always kept an eye out for the first saguaros. They were to be found just below a plateau edge, a little more than 3000 ft. elevation, protected by a basaltic rim from chill winds, but warmed by heated air rising from the desert valley below. A few hundred yards away, on the flats above the rim, not a saguaro is to be found. The growth of organ pipe cacti at the northern limit of their range is also limited by cold. The plants often exhibit constrictions in the trunks and arms, where freezes have caused the death of meristematic (growth) tissue, at times causing the branches to resemble stacked green knobs.[22] (Fig. 1.24.)

In tropical deciduous forests (seasonally dry tropical forests), the green trunks and branches and other special drought-coping mechanisms allow columnars to continue to be physiologically active and grow during spring and fall (and often winter) droughts, whereas leaf-bearing trees drop their leaves and are forced into relative inactivity by the sustained drought or high temperatures or both. In the late spring, the remainder of the forest may appear dead or dormant, but close examination reveals the great cacti to be covered with buds, flowers, or fruits. Hillsides of thornscrub or dry tropical forests resemble postfire wastelands except for the greens of columnars—etchos, pitayas, sahuiras, and in Ecuador cardones *(Armatocereus cartwrightianus* and *Pilosocereus tweedyanus).*

The living requirements for each species are often varied. Saguaros cannot survive on flatlands at the northern and southern extremes of their distribution, but do rather well on the flats in between. The explanation for this ecological differentiation is that at the northern limits the lowlands are often chilled with a layer of freezing air, while the hillsides are warmed with upwelling heated air from below that fends off damage by freezing. Saguaros above the flats stand much better odds of surviving the winter. At the southern limits, steep basaltic hillsides where the last saguaros survive are hotter and drier than the surrounding countryside. Sagua-

Figure 1.24. Organ pipe (Stenocereus thurberi) *with frost-produced constrictions, Organ Pipe Cactus National Monument, Arizona.*

Figure 1.25. Trichocereus atacamensis *with heavy coat of hairs as frost protection, Tastil Archaeological Site, Salta, Argentina.*

ros tolerate the heat and drying and thus outcompete the numerous tropical competitors that strangle or crowd out young saguaros elsewhere. Pasacanas generally flourish in the cool higher Andes above 1500 m (5000 ft.). They grow in their greatest densities at higher than 3000 m in places such as the profound Quebrada del Toro in Salta province, Argentina. At that elevation, they are protected by a denser coat of spines, especially at the growing tips, thicker and more abundant than seen on pasacanas growing at lower elevations. Scattered woolly individuals can be found at near 4000 m in southwestern Bolivia, where temperatures fall as low as -10°C (14°F). (Fig. 1.25.)

The range of rainfall in which columnar cacti are to be found is vast. In parts of its homeland, the odd *Browningia candelaris* of northern Chile and southern Peru appears to be the only plant in its habitat, growing on barren slopes where rain may not fall for years. Rain may never fall on the fog-loving *copao* (*Eulychnia* spp.) from northern Chile. The huge sahuesos achieve their greatest growth at between 75 and 150 mm (3–6 in.) of rainfall, a point at which saguaros tend to be unable to survive. In contrast, the rare *Neobuxbaumia sanchezmejoradae* grows only on craggy and eroded limestone outcroppings rock in a jungly region of southwestern Oaxaca that receives about 2500 mm annual

rainfall, more than 100 inches.[23] The tall and graceful *Espostoa utcubambensis* of the Amazonian tributary Río Utcubamba in the Peruvian Andes thrives on steep hillsides in similar wetness. The large genus *Cereus* obtains its maximum diversity in Brazil and environs where precipitation in excess of 1000 mm is commonplace. (Fig. 1.26.)

How long do columnar cacti live? The answer is that many live for a long time. Dating the age of cacti is difficult, though, because they do not leave annual growth rings, as do coniferous and deciduous trees from temperate climes. Saguaros are known to live at least two hundred years, perhaps more.[24] Ecologist Ray Turner suggests that sahuesos live as old or older, perhaps as much as three hundred years.[25] Pasacanas probably live at least two centuries.[26] Mayos inform me that the average adult pitaya is at least one hundred years old, a figure they derive from oral tradition and continuous observation of individual plants across generations. Most columnars in very arid places (less than 250 mm [10 in.] of annual rainfall) grow slowly indeed. Many (but not all) species, even those in areas with warmer year-round climate and annual rainfall in excess of 500 mm, need fifty years to reach adult size.[27] Cuttings planted in the ground may yield fruit after only three or four years, but these plants can hardly be considered adult size.

Figure 1.26. Browningia candelaris *on hillside, Arequipa, Peru. Rainfall is under 100 mm (4 in.).*

Distribution of the Great Cacti

Although most of us North Americans think of columnar cacti as growing in the American Southwest and throughout Mexico, they are also important in the vegetation of South America—including some tropical forests—and the Caribbean. Every South American country hosts at least one species of columnar cactus. The Florida Keys are host to an endangered skinny columnar, *Pilosocereus polygonus*, Florida Key tree cactus, as are the Bahamas. Some of the densest forests of columnar cacti are found on the islands of the Netherlands Antilles. From the high Andes to the Brazilian sertão to the desert of Aruba and the lushness of Cuba, the giant cacti grow and produce fruit and lumber. Horticulturalists of the pre-Incan Dieguito Culture of the high, dry Andes cultivated *Trichocereus atacamensis* for food and lumber. Numerous species of the genus *Cereus* grow wild in humid South America east of the Andes. (Fig. 1.27.)

The species of columnars that flourish in the moist tropics *(Cereus, Espostoa, Neobuxbaumia euphorbioides)* tend to favor rocky substrates where potentially competing species cannot establish roots or seasonally dry tropical forests where the months of aridity render their deciduous competitors for sunlight both leafless and relatively dormant. During the time of little or no rain, the cacti grow in spurts and usually flower and fruit as well. The dry season is often the only practical time to search for and study the great columnars of the dry tropical forests such as etchos, *gigantes (Neobuxbaumia mezcalaensis),* and *pitayales (Stenocereus chacalapensis).* I happened upon an undescribed species of *Neobuxbaumia* one day on the coast of the Mexican state of Oaxaca, but the surrounding trees and shrubs were so dense and thorny that I could hardly get close enough to examine and photograph it. Had I been passing by during the lushness of the rainy season, I would not have spied the plant. (Figs. 1.28 and 1.29.)

Figure 1.27. Forest of yatos (Stenocereus griseus), *Bonaire.*

Figure 1.28. Gigantes (Neobuxbaumia mezcalaensis)
near Izúcar de Matamoros, Puebla.

Figure 1.29. Undescribed Neobuxbaumia,
Huatulco, coast of Oaxaca.

Species of columnar cacti number more than 130, and all are native to the New World (including the West Indies), as are all members of the family Cactaceae. The genus *Rhipsalis,* an epiphytic cactus, may be native to tropical Africa as well as to the New World, but it was more likely introduced by Europeans or birds. Its seeds are very sticky and could easily have adhered to birds, mammals, people, or baggage traveling from Brazil to East Africa, where it is now found. Cacti seen growing elsewhere in the world are transplants from the Western Hemisphere. The cactus pioneers N. L. Britton and J. N. Rose described most of them in their remarkable 1920 work *The Cactaceae.* In the mid–twentieth century, the German cactologist Curt Backeberg noted forty-seven species of columnar cacti with large white flowers in Argentina, Bolivia, Chile, Ecuador, and Peru *alone.* (He was well known as a taxonomic *splitter,* one eager to name new species perhaps without sufficient allowance for variance within a species.) Dennis Cornejo identified fifty-nine species of columnars in Mexico in 1994. Today that number has risen to at least sixty-two, probably more.[28] I say "at least" because the identification of several species is controversial, and taxonomic work remains to be done to determine if they are distinct species. Two new species have been described since Cornejo's study, and at least two undescribed species puts the number of Mexican columnars in the sixties. Brazil may host more than forty species in the genera *Cereus, Pilosocereus,* and *Facheiroa* alone, and though most of them are of small stature, Brazil may surpass Mexico in species of columnars.[29]

In at least three major centers of cactus diversity, especially large numbers of cactus species evolved: Greater Mexico, including Arizona, California, and Texas; the Andes, especially Peru, Bolivia, Argentina, and Chile; and eastern Brazil. Each location has its own salient tribes (a taxonomic classification between the subfamily and the genus) and genera. Each has its range of identifying ecological conditions,[30] and within these centers are specific "hot spots" where columnar cacti species proliferate. (See chapter 3 for an overview of hot spots and their species.) The most remarkable hot spot is the complex of the Valles de Tehuacán and Cuicatlán in the south central Mexican states of Puebla and Oaxaca. This network of arid valleys descending eastward into the Papaloápan basin from the northwest and south is not only an explosive center of cactus diversity, but also has seen human habitation for nearly ten thousand years. The warm (not terribly hot) valleys are home to nine or ten genera (depending on one's taxonomic biases) and at least eighteen species of columnar cacti, some of them—tetechos, viejitos, and chicos—in dazzling numbers. Hillside species are usually distinct from those that frequent the valley floors. There are three species of *Neobuxbaumia,* four of *Pachycereus,* two of *Myrtillocactus,* two of *Polaskia,* and three or four of *Stenocereus.* At least ten columnars in the Valle de Tehuacán are of considerable economic importance to the inhabitants of these communities, and all species are (or formerly were) used in one way or another. At least seven species are cultivated and should be considered to be semidomesticated. Thus, the valley is a center of diversity, both in the proliferation of cactus species and in their ethnobotanical applications—that is, the variety of human uses. (Figs. 1.30 and 1.31.)

Slightly less diversity but more endemism is found in the Río Balsas and Tepalcatepec basin (or depression) in the southwest Mexican states of Guerrero and Michoacán, especially in the environs of Presa Infiernillo in Michoacán, a reservoir that forms the state line with Guerrero. The rugged topography and torrid climate kept much of this area sparsely populated until the 1950s, so the archaeological record is less developed, and ethnobotanical uses of the cacti have not been comprehensively cataloged. An even less-studied center of diversity in Mexico is in the region of Totolapan in the rugged canyon country of eastern Oaxaca. Within a 40 km (24 mi.) radius, at least five endemic species or varieties are to be found, some of them of most unusual appearance. (Fig. 1.32.)

In some South American centers, diversity is diffused over a large area and, with one exception, is not concentrated in small regions as in Mexico. The exception is the long Río Marañon valley of Peru that harbors a wide variety of habitats and a surprising diversity of columnars, including the seven genera. Brazil, especially the large area occupied by the states of Bahia, Goais, and Minas Gerais, hosts remarkable proliferation of the genera *Pilosocereus,* the bearded cacti (perhaps more than twenty-five species); *Cereus* (three species); and *Facheiroa* (at least eight species). Most of these species are of highly limited distribution, often being found only on specific substrates and with tiny ranges, and some are of questionable columnarhood as well. In the central highlands of Bolivia and the high Puna of northwest Argentina and Bolivia, there are numerous fine and rather dense stands of columnars as well. Even in Ecuador, with its generally moist to wet climate, small arid belts in the south in the vicinity of Catamayo, west of Loja and in the valley below Santa Isabel west of Cuenca, are home to several columnar cacti. On the southwestern coast, columnar cacti are numerous. Madsen notes thirty-five species and fourteen genera of Cactaceae in the country.[31]

Southern through northwestern Mexico seems to be the ideal host for columnar cacti. In the semiarid Río Mayo region of northwest Mexico, an area less than 10 percent

Figure 1.30. Chicos (Pachycereus weberi), *Calipan, Puebla. Note the absence of small plants.*

Figure 1.31. Diverse cactus forest near Zinacatepec, Valle de Tehuacán, Puebla. Included are Stenocereus stellatus, S. pruinosus, Escontria chiotilla, Myrtillocactus geometrizans, *and* Pachycereus weberi.

the size of Ecuador is home to at least fifty species of cacti in thirteen genera, four of them including columnar cacti. Valle de Tehuacán has more than twice that number in a smaller area. Mexico, with a land area five times that of Ecuador, has more than nine hundred species of cacti, and cacti are prominent among plants used by people. Given the immense variety of usable plants in Ecuador and the tiny proportion that are cacti, it is not surprising that human uses of products derived from the humble cactus are nearly nonexistent there. (Fig. 1.33.)

The greatest diversity of columnars and greatest number of large cacti grow in arid to semiarid valleys of central

Figure 1.32. Pitires (Stenocereus quevedonis) *near Presa Infiernillo, Michoacán. Note the heavy overgrazing and scarcity of smaller plants.*

Figure 1.33. Pilosocereus tweedyanus *in tropical deciduous forest, southern coastal foothills, Ecuador.*

Mexico, including the Balsas basin and Cañon Infiernillo (Guerrero and Michoacán), the Barranca de Metztitlán (Hidalgo), Valle de Tehuacán (Puebla), and Valle de Tomellín (Oaxaca). The proliferation of species of columnars—at least twenty-five—in the canyons and narrow valleys of Guerrero, Michoacán, and Puebla is truly astounding. In these states, the densest and most diverse growths have been spared the machete and plow that have beleaguered plant communities elsewhere (especially in Brazil). In the United States, columnars grow only in Arizona and Florida, with a few token saguaros residing in California near the Colorado River.

Columnar cacti in North America tend to adhere to what ecologists refer to as Rapaport's Rule—namely, that the north-south range of cacti tends to decrease toward the tropics and increase toward the temperate zones.[32] Thus, etchos have the greatest north-south distribution—over 2000 km (1200 mi.), extending well into the temperate zone and the tropics. The columnars with the most extensive latitudinal distributions (in this case northwest to southeast) are located primarily in the subtropics. In addition to etchos, they include saguaros (1000 km), sahuesos (1000 km), and organ pipes (1200 km). At the same time, species of the Valle de Tehuacán in Puebla, all tropical (but with moderate elevational sorting), tend to have far narrower distributions. Indeed, the range of four species there is limited to a 100 km (62 mi.) radius or less.[33] The same is true of at least three species found in the vicinity of Presa Infiernillo.[34] Rapaport's Rule does not apply as neatly in South America, perhaps due to the dramatic elevational gradients produced by the Andes that diminish the effects of the tropics. Even so, pasacanas of the high, cold Andean valleys have a significantly greater latitudinal distribution than any of the more tropically inclined columnars growing to the north. Three of Ecuador's seven species of columnars, growing just south of the equator, are endemic to a small region of less than 50 km radius in the southwestern part of the country.[35] The Río Marañon *west* of the Cordillera Oriental of Peru, in an area only 8° or so south of the equator, is home to at least twelve narrowly endemic species.[36] The very tropically oriented *Neobuxbaumia sanchezmejoradae, N. multiareolata, Stenocereus zopilotensis,* and *S. chacalapensis* of southern Mexico may have a range of only a few kilometers. The same is true of several members of the genus *Pilosocereus* in Brazil. (Fig. 1.34.)

Columnar cacti grow much farther north on the Pacific side of the North American continent than on the Gulf of Mexico side. Saguaros, for example, seem to prosper on south-facing mountainsides at roughly 35°N latitude in Arizona, but no columnar cactus is to be found north of

Figure 1.34. Stenocereus chacalapensis, *Oaxaca coast.*

22°N latitude in Tamaulipas state and not at all in Texas, even though its southernmost point lies at 26°N latitude. The absence of columnar cacti in this region is due to the occasional arrival of powerful hurricanes that would surely topple or drown the tall plants and to the rarer "blue northers" or frigid masses of Arctic air that occasionally sweep down the Great Plains and may produce hard freezes as far south as southern Tamaulipas. Northwestern Arizona, where the northernmost saguaros grow, is free of strong hurricanes and sustained freezing. As we shall see, columnar cacti in general are intolerant of sustained freezes and find the blue northers abhorrent. In South America, the cold-tolerant *Trichocereus atacamensis* reaches roughly 28°S latitude on the east slope of the Andes, and *T. chilensis* reaches somewhat farther south on the west side.

By far the most cosmopolitan—and prolific—genus is *Pilosocereus,* usually a bearded cactus. Somewhat more than thirty-six species embrace a variety of habitats from southern Brazil to Paraguay north through Mexico to within 350 km (210 mi.) of the U.S. boundary, both in Sonora and in Tamaulipas (different species), and into southern Florida and the Caribbean. *Pilosocereus* grows in montane oak forests, tropical deciduous forests, thorn-scrub, and semideciduous tropical forests. It does rather well in both desert and moist tropical climates on islands of the Caribbean and is native to the Florida Keys. Only a few species can be viewed as cosmopolitan (the range of *Pilosocereus alensis* extends from Sonora southeastward for more than 1000 km). The majority of species in this genus have a distribution that is confined to a small area. One Brazilian species, *P. magnificus,* appears to be limited to a radius of no more than 10 km (6 mi.), perhaps less. The various species are often difficult to distinguish, and additional species will undoubtedly be added to the list—for example, the seemingly undescribed species shown in figure 1.35. (Fig. 1.35.)

Spaniards arriving in the New World—and all subsequent visitors with an eye for plants—found a bewildering variety of cacti. Europe had no plants that even remotely resembled cacti, so they were objects of intense curiosity. Some travelers, however, had undoubtedly seen a smattering of arborescent African *Euphorbias,* some of whose shapes bear an uncanny resemblance to cacti. I recall my fixation on cacti and the thrill of seeing them for the first time in their habitat. I had seen them in photographs. The early Europeans probably hadn't even seen them in drawings! When they first encountered prickly pears, they immediately shipped seeds and plants back and planted them. Prickly pears ("cactus pears" in Europe) have become an important agricultural product in the Mediterranean, and pitahayas, fruits of the spreading, vinelike *Hylocereus undatus* that grows in trees and on roofs in lowland Mexico and the Caribbean, are enormously popular in Southeast Asia. (Fig. 1.36.)

Spaniards (northern Europeans for the most part did not see cacti until much later) christened these strange, treelike plants *cardones,* which translated simply means "big thistles," a most unimaginative title. As a result, dozens of species of cereoids are graced with the uncreative name *cardón,* even though the differences among the various cardones are spectacular. Somewhat less common but still widespread is the Spanish name *órgano,* or organ, applied to at least five species of Mexican cacti. Also common in North America is the name *pitaya* and some variations, apparently derived from a Caribbean term, but now mostly confined to the plants and fruits of the genus *Stenocereus.* Some columnars in North and South America are also christened *gigantes* (giants). In South America, several species are simply called *cactus.* The Sonoran Desert region was blessed with the retention of indigenous names more than in any other region. We have etchos, saguaros, sahuesos, sahuiras, *sinas (Stenocereus alamosensis),* as well as the

Figure 1.35. Pilosocereus *sp. near Cintalapa, Chiapas.*

indigenous *aaqui* (Mayo for pitaya) and *matagochi* (Guarijío for *Pilosocereus alensis*). The Seris retain an indigenous name for every cactus found in their region.

Some early European observers described the plants in detail, fully aware that their audiences in the Old World would be incredulous. The Jesuit chronicler Andrés Pérez de Ribas, writing in 1645, described the organ pipe, which he found growing in the deserts of northwest Mexico in great numbers. (Fig. 1.37.)

Most of the year they use the pitahaya [pitaya], which grows abundantly here. It is an unusual plant in European terms or even compared to other trees in the world. Its branches are like green grooved candles growing straight up from the trunk, which grows short and robust, making a very striking crown. It has not a single leaf, but the fruit appears on the branches like nipples. Because of its thorny surface, this fruit looks something like the chestnut or the prickly pear [fruit]. Its flesh is similar to that of the fig, but softer; the flesh is very white in some plants, red or yellow in others. It is very delicious, particularly when grown

Figure 1.36. Pitahaya,
Hylocereus undatus,
*Puebla. This cactus
has become popular in
Southeast Asia, where
the fruits are known as
dragon fruits.*

Figure 1.37. El Pitayal, forest of pitayas (Stenocereus thurberi), *southern Sonora. More than half of this forest has been
cleared for agribusiness and shrimp farms.*

in the dry soil along the seashore in the Province of Sinaloa [southern Sonora], where it rains very little. These trees are so abundant that there are groves that stretch for two, three, and up to six leagues.[37]

The Jesuit Ignaz Pfefferkorn, writing more than one hundred years later, following the Jesuit expulsion from the Spanish colonies, described the organ pipe as follows:

The pitahaya bush grows principally on knolls and hills although it is also found on flat land. As for size, it could well be called a tree, since it sometimes reaches a thickness of more than two feet and attains a height of ten to twelve or thirteen ells. Frequently it has ten, twelve, and even more branches, each of which is as thick and as long as the main trunk. Both branches and trunk have longitudinal furrows all around like melons. All are a light green color and are covered everywhere with pointed spines, which, however, the tree gradually loses as it ages. On this admirable shrub grows the sweet, pleasant-tasting pitahaya. It does not hang on small stems like other fruits, but clings directly to the trunk and the branches. Like the rest of the tree, it is entirely covered with spines. In fruitful years there are often several hundred on one shrub. They are in size and shape like a duck egg with thick, tough, green shell, which becomes light brown when ripe. The meat of the pitahaya is either white or blood red and is full of black, easily chewed seeds like those found in figs. Red pitahayas are the most plentiful in Sonora; white ones are somewhat scarce. Many Indians open this fruit, lay it in the sun for a day to dry up the abundant juice. They then loosen the meat from the shell and press it very hard into a kind of cake, which is called a tamal. The cakes last a long time without spoiling and are relished by the common man.[38]

Most columnars have a flowering and fruiting sequence judiciously coinciding with the onset of warm-season rains. Thus, their seeds spill onto the ground from fallen (or digested) fruits at the same time that rain is most likely to fall in hot weather, assuring the plant optimum opportunity for reproductive success. Pasacanas, pitayas, saguaros, sahuesos,[39] sahuiras, and tetechos all demonstrate this pattern, each exhibiting a different fruiting time, a sure indicator that maximum probability of rain in their peculiar habitat cannot be far behind. *Pitaya agria (Stenocereus gummosus),* a shrubby, semicolumnar of Baja California (a few plants grow on the Sonoran coast), flowers and fruits in the fall, apparently in preparation for the possibility of episodic heavy rainfall from Pacific hurricanes or the more

Figure 1.38. Pitaya, perhaps undescribed Stenocereus, *Isthmus of Tehuantepec.*

gentle soaking of winter rains. Sufficient conditions for setting of seed, germination, and survival of these species may occur only every decade or so, but such is the longevity of each plant that the investment and risk in fruiting is worth the gamble. The gigante (*Neobuxbaumia mezcalaensis*) of south-central Mexico flowers and fruits in late May and may continue well into September, as is the case with the *xoconochtli (Stenocereus stellatus)*. How this flowering and fruiting sequence favors selection and ties in with climate needs field research.

After pondering the distribution and densities of columnar cacti, I have come to the conclusion that many species, not just the few that are cultivated, have been deeply intertwined with human beings, who, in turn, have been knowing (and perhaps unknowing) agents for the plants' increased distribution. The wide and increasing range of *Cereus repandus, C. jamacaru, Pachycereus pringlei* (in Sonora, Mexico), *P. marginatus, Stenocereus gummosus,* and *S. griseus* point directly to human intervention. The latter species is so widespread and heavily cultivated that the origin of the aboriginal plants is difficult to pinpoint. Humans are heavily implicated in the distribution of at least ten additional species and probably more, especially species of the genus *Stenocereus,* which are the best producers of edible fruits. (Fig. 1.38.)

What Makes a Cactus?

For those people whose exposure to plants is limited to the plants that grow in more temperate or wet climates, cacti are decidedly strange. In addition to numerous American collectors, many Europeans and Asians find them intriguing and charismatic. Collectors and horticulturalists nurture and propagate them by the millions. In addition to arborescent and columnar cacti, members of the cactus family range from the barely noticeable Andean genus *Blossfeldia*, weighing less than an ounce or two, to shrubby plants several meters high and covering several acres *(Stenocereus gummosus)*, to vines curling snakelike through trees *(Hylocereus)*, to epiphytic masses with roots that never touch the ground *(Rhipsalis, Selenicereus)*, and to huge trees weighing many tons *(Armatocereus, Backebergia, Browningia, Carnegiea, Cephalocereus, Cereus, Corryocactus, Espostoa, Mitrocereus, Neobuxbaumia, Neoraimondia, Pachycereus, Stenocereus,* and *Trichocereus,* to name a few). (Fig. 1.39.)

What all cacti have in common, what distinguishes members of the cactus family (Cactaceae) from other families, includes a combination of characteristics. First, as adults they all employ the Crassulacean acid metabolism (CAM) pathway of photosynthesis to produce new growth. CAM, found in as many as twenty plant families, sets cacti and other succulents apart quite simply in that they open their pores (stomata) and carry on their uptake of carbon dioxide at night, when other plants have shut down their pores to sleep, as it were. By closing their stomata during the day and opening them at night, CAM plants hoard precious water vapor, trapping it *inside* during the hottest hours. At night, they store carbon dioxide and wait for sunlight before they undertake the complex biochemical processes that produce growth, quite a remarkable accomplishment. Without CAM, cacti could never have adapted with such marvelous and varied success to their hot, arid habitats and have assumed their wide assortment of shapes and sizes.CAM plants lose less than a third as much water when their stomata are open as plants with non-CAM metabolism.[40]

And, of course, cacti have modified leaves, more familiarly known as *spines,* that emerge from special structures called *areoles,* wartlike lateral buds. Most areoles have a cluster of spines ranging from one to many. The areole/spine anatomy sets cacti off from other plants. Spines serve critically important functions in addition to their obvious deterrent value. The absence of true, flat leaves, except in primitive cacti such as *Pereskias* and *Quiabentias,* and of small, ephemeral, pointy leaves, except in various *Opuntias,* permits the cacti drastically to cut back on water loss and

Figure 1.39. Saguaro seedlings, Bach's Nursery, Tucson. More than one million young plants are represented in this photograph.

leaf wilt and thus to protect themselves from damage due to high temperatures or extreme drought. The leaf-type origin of spines (as opposed to the branch-type origin of thorns) can best be seen in some species of *Opuntia,* wherein the developing leaves are succulent and often edible (they serve as a snack food in some regions), then drop off with age. Spines function as axillary buds, gaining their rigidity and vicious function with age. Spines occur in marvelous variety—setose (bristlelike), needlelike, subulate (tapered to a sharp point), conical, cylindrical, planed, straight, curved, twisted, hooked, and feathered—to name the most common.[41] They may point outward, upward, sideways, or downward. For example, each areole of *Pachycereus hollianus,* the *baboso* of the Valle de Tehuacán, has one long downward-facing spine. It is a most effective defender against herbivores and other intruders and has led to the plant's use in living fences for corrals. (Figs. 1.40, 1.41, 1.42, 1.43, and 1.44.)

Spines range in length from 1 mm (1/25 in.) to more than 30 cm (12 in.) and in a range of colors from opaque to maroon, purple, red, orange, black, and white. The longest

Figure 1.40. Well-armed areoles of Stenocereus montanus. *Note the single long central spine.*

Figure 1.41. Areoles of Neobuxbaumia scoparia. *Areoles are closer together and spines less numerous and shorter than in* S. montanus.

Figure 1.42. Areoles of Stenocereus friçii. *Note the irregular ridge of the ribs and the numerous, even-lengthed spines.*

Figure 1.43. Areoles of Neoraimondia arequipensis. *Areoles are massive and felty. Spines are few, but very long.*

Figure 1.44. Areoles on Trichocereus terscheckii, *Salta, Argentina.*

Figure 1.45. Fruits of Pachycereus hollianus, *Valle de Tehuacán, Puebla.*

ones tend to grow on the high-elevation columnar cacti in the Andes, where they serve as insulators from cold and ultraviolet rays as well as defenders of the plant. Spines may be nearly as thick at the base and as rigid and long as ice picks. In contrast, in the case of *Opuntias* and some primitive cacti such as *Pereskiopsis,* the cacti sport tiny, barely visible spines called *glochids* that are covered by a removable sheath (often itself armed with reverse barbs and a toxin) that tends to remain embedded in the flesh when the remainder of the spine is removed, a more lasting reminder not to mess with the plant. Lovers of columnar cacti can rejoice in the fact that these plants—and their fruits—are glochid free. (Fig. 1.45.)

Spines provide cacti with camouflage and an effective defense against large herbivores. Even more important in some cases, the thick growth of spines and areolar hairs at and near the growth tips *(apical meristem)* provides protection for the cactus against freezing and sunburning. The numerous spines at the growth tips of saguaros allow them to grow considerably north of the freeze limits. The minimum temperatures at the growing point are increased by as much as 6°C (10°F) and

decreased from high temperatures by as much as 11°C (18°F) by the presence of spines.[42] These same spines can also, in some species, act to shade the plant from intense sunlight. In climates where fog or heavy dew is common, spines direct water to the plant and downwards to the ribs and to the roots. Spines are often indirectly involved in propagation. Some *Cylindropuntias* (chollas) may release a detachable cladode (spiny segment) so easily that it "grabs" at passing mammals, detaches, hitchhikes for a while, and, finally, falling to the ground well away from its parent, takes root and assists in propagation of the species.

In addition to spines, several genera produce small or rudimentary true leaves, but only a few primitive species (*Pereskia* and *Pereskiopsis,* for example) have true or fully developed leaves. At first glimpse, a *Pereskiopsis* looks more like a snaky green vine with odd leaves than it does a cactus. However, those unfortunate enough to grab the stem will discover that the plant is laden with hundreds of thousands of potent glochids, as if to prove that it is indeed a cactus. Some *Pereskias* and *Quiabentias* resemble weak-stemmed trees, whereas others at first blush appear to be simple

Figure 1.46. Pereskia, *probably* bahiensis, *Río San Francisco, Bahia, Brazil. In spite of its abundant leaves, the plant is still a cactus, albeit a primitive one.*

thick-trunked, wide-crowned trees, looking not at all like a cactus. (Fig. 1.46.)

In most cacti (and in all columnar cacti), the clusters of spines and flowers and fruits grow out of areoles. The areoles may provide the distinguishing character that enables the observer to differentiate among numerous bewilderingly similar species. Areoles often resemble tiny pincushions with the pins placed pointing outward. They may be raised, resembling a knob *(Neoraimondia)*, or they may be flat or even slightly concave *(Azureocereus)*. In cereoids, they are located on the ridge of the ribs and give rise to hairs and spines of varying lengths. New growth—flowers, branches, and spines—emerges from the areoles, and the pattern of growth is often distinct for each species. The number, shape, size, and color of areoles; the distance between them; the size, shape, and contour of the ribs on which they grow; as well as the number, length, shape, size, and arrangement of the hairs and spines that grow from

them can be as diagnostic of the various species as fingerprints are for individual humans. The contrast among just a few species is a microcosm of the enormous variation of areolar characteristics among columnar cacti.

All columnar cacti have vertical ribs arranged rather like the baffles of an accordion. This construction permits the plant to expand as it absorbs water—often large amounts—in rainy times and to shrink dramatically during dry times without damaging the plant or causing tissues to burst. The stems abound in tissue called *spongy parenchyma,* which is composed of cells capable of vast expansions, making the plants resemble living storage tanks that can release the liquid when the plant requires it. The parenchyma expands as it takes in water, causing the ribs to balloon as if inflated. In deserts, the rapid change in the shape of a columnar cactus following a heavy rain can be dramatic: overnight a skinny, emaciated saguaro can be transformed into a fat, complacent giant.

Cacti, especially many columnars, also exhibit in death a characteristic woody skeleton. The succulent flesh usually decays and falls off the tougher wood of the trunk rather soon after actual death; that is, the green epidermis and the tissues beneath soon turn brown and yellow, then decompose, leaving the mostly hollow skeleton behind to disintegrate at its leisure. Because so many cacti grow only in desert environments where moisture is low, plants often do not rot as they do in wetter climates, and these noticeable, sometimes large, skeletons may persist for years, often upright. Columnars usually have more pronounced skeletons than lower-growing cacti because they require a strong structure to hold up their heavy, water-laden flesh. A 12-m-tall (40-foot) sahueso weighing 25 tons needs a durable and rigid internal arrangement simply to remain upright.

An additional characteristic of cacti, except for the primitive genus *Pereskia,* is that their flowers exhibit inferior ovaries; that is, the ovaries are located below the attachment point of the stamens (male sexual parts)—a characteristic they share with numerous other plant families. Some researchers theorize that this somewhat buried position of the sexual organs affords greater protection against nibbling, chewing, and sucking herbivores than does a superior ovary arrangement. Inferior ovaries indicate a highly specialized group of plants.[43]

Healthy cacti usually have green or greenish bark or skin, enabling them to carry on photosynthesis on nearly all exposed surfaces. In some columnars, the trunks (and branches in the fog-obligate genus *Eulychnia*) develop a gray bark similar to that of other trees, but the exterior of the branches usually remains green. Because only a few primitive cacti support leaves, the green bark enables cacti to manufacture the sugars that allow them to grow. The absence of leaves also makes it essential for cacti to obtain maximum exposure to sunlight in order to carry on photosynthesis. For this reason, cacti tend to be absent in plant communities where rainfall is high and vegetation is usually luxuriant. Where they do survive in areas of high rainfall (greater than 1000 mm, or 40 in., annually), they tend to be located on rock outcrops, thus poised to intercept maximum available sunlight and avoid competition with other trees that might block the sun's rays. The rocky surface also assures that rain will run off and not produce waterlogged soils that would induce rotting in the cactus roots. The green stems and branches expose a large surface area to sunlight, making the giant plants highly effective solar panels. The largest saguaros tend to grow near the northern extremes of distribution, places where sunlight is least direct and cold is a limiting factor. The greatly increased surface area of the larger cactus renders them capable of absorbing much more sunlight than small plants, and the increased ratio of surface area to mass may assist in combating freeze damage. In the case of *Trichocereus atacamensis,* however, the plants at the highest elevations are for the most part smaller than those at the lower limits. These higher-elevation plants devote much of their growth into covering themselves with a dense coat of spines that protect against freezing and probably shield against the excessive ultraviolet rays of the thin atmosphere of the high Andes.

Finally, nearly all cacti have clear sap or gum. When a cactus is traumatized, the cuticle or epidermis may turn different colors or demonstrate a variety of responses, often aiding in identification, but the liquid that runs or oozes is clear.

These combined characteristics—CAM metabolism, spines, areoles, woody skeleton, inferior ovaries, green bark or cuticle, and (for the most part) clear sap—set off cacti from other families. Many Old World members of the genus *Euphorbia* superficially resemble cacti (in South Africa they are even called *cactus*); they may have spines, green bark, and the characteristic shape of cacti—a classic case of convergent evolution. The spines of these plants, however, do not grow from areoles; the sap is (often) milky; and all have leaves, many of which pop out only during and following rainy seasons.

Species of cacti (the family Cactaceae) number between about eight hundred and two thousand depending on the authority one wishes to cite.[44] Siding with either extreme renders one vulnerable to angry attacks by proponents of the other extreme. Splitters (taxonomists who accentuate differences) consider lumpers (taxonomists who point to similarities) muddle-headed, while lumpers view splitters as simple-minded. As J. D. Mabberly, a well-known plant taxonomist notes, "The number of genera is perhaps still inflated by overfamiliarization in horticulture and over-stressing of trivial features conspicuous therein."[45] The agreed-upon number will change as more field research clarifies distribution of the columnar cacti and laboratory work hones in on genetic relationships. Anderson, following the International Cactaceae Systematics Group, recognizes 125 genera and 1810 species.[46] Well more than 130 of these species are columnar cacti, with at least 60, probably more, of the great ones in Mexico alone. So the number of cactus species matches that of many other well-known plant families. (Figs. 1.47, 1.48, and 1.49.)

The great cacti assume a wide variety of shapes. They may be a single stalk, straight as a telephone pole (*Neobuxbaumia mezcalaensis, N. polylopha, Pachycereus margin-*

Figure 1.47. Espostoa frutescens, *near Santa Isabel, Ecuador. I consider it a columnar cactus in spite of its short stature.*

Figure 1.49. Sina (Stenocereus alamosensis) *in flower. The tubular flowers are hummingbird pollinated. Its branches are too thin to qualify it as a columnar cactus.*

Figure 1.48. Stenocereus gummosus, *Baja California. This cactus is typically a rambling plant. Some grow tall enough to warrant consideration as columnar cacti.*

atus), or slightly inclined, like ranks of stooped-shouldered soldiers *(Cephalocereus columna-trajani)*, or a tangle of intertwined branches (Cornejo's *nonerect* columnars), or a mishmash of these forms *(Cereus jamacaru, Stenocereus alamosensis)*. They may be shaped like a swimming squid *(Stenocereus thurberi)*, an upside-down daiquiri glass *(Escontria chiotilla)*, a goblet *(Pachycereus weberi)*, a hooded shepherd *(Trichocereus atacamensis* at high elevations), or just simply a great cactus. (Fig. 1.50.)

The great size of some columnars suggests that they should be considered trees. James Ohio Pattie in 1825 wrote in his diary on seeing his first saguaro: "a species of tree.... It grows to a height of 40 or 50 feet."[47] A fine example of a columnar cactus that is a tree can be seen in figure 1.34.

Their role in vegetation also varies dramatically both between and within species. In places, they grow together in great numbers resembling forests *(Browningia pilleifera, Carnegiea gigantea, Cephalocereus columna-trajani, Cereus repandus, Espostoa utcubambensis, Myrtillocactus geometrizans, Neobuxbaumia tetetzo, Pachycereus pringlei, Stenocereus thurberi)*, or as isolated loners *(Carnegiea gigantea, Browningia* sp., *Mitrocereus fulviceps, Pachycereus [Lophocereus] schottii)*. The loners may be genetically programmed to eschew companionship, or they may be genetic lovers of companionship that are merely incidental outliers and in the appropriate conditions would form forests or semiforests.

Ribs, Flowers, and Roots

Among columnars, the rib count varies between three and about thirty. That number remains more or less constant within most species and is often a key in distinguishing similar species. The comparative numbers of ribs can also be a key to the evolutionary relationships among different species.[48] Ribs also have different shapes (rounded versus sharp, smooth versus irregular), depth, and prominence.

The flowers of columnar cacti range in aesthetic quality from delicately pretty to voluptuous. As a rule, the flowers are large (nearly all in excess of 25 mm [1 in.] in diameter) and are often huge—up to 20 cm (8 in.) in diameter with a 20-cm floral tube. The dominant shade is white, but colors are equally varied, ranging from drab green *(Pachycereus gaumeri)* to delicate yellow *(Escontria)* to shocking pink *(Stenocereus stellatus)* to reddish pink *(Neobuxbaumia euphorbioides)* to red *(Neobuxbaumia polylopha)* to purple *(Neobuxbaumia multiareolata)* to black and white *(Azureocereus, Neobuxbaumia mezcalaensis)*. White or delicately tinted flowers usually open at night, presumably to be more visible to bats or hummingbird-moths (hawk moths).

Figure 1.50. Chico (Pachycereus weberi), *Cañon de Tomellín, Oaxaca. Fruits are tasty, but the rinds are tough.*

Strongly tinted flowers usually open during the day and are pollinated by insects or hummingbirds. (Figs. 1.51, 1.52, 1.53, and 1.54.)

In some species, all or nearly all flowers open simultaneously, whereas in others they open in a staggered fashion over several weeks to more than a month. A few species flower intermittently throughout the year, some opportunistically (whenever they feel like it), others in regular flowering seasons. In my front yard, a Peruvian apple cactus *(Cereus hildmannianus)* one night exhibited more than forty flowers opening at the same time, all with a diameter of 20 cm (8 in.). By midmorning the following day, all had closed and drooped. The plant usually buds and flowers about two weeks following any warm-season rain. Most cactus flowers endure less than twenty-four hours, much to the disappointment of flower watchers. This is apparently a water-saving strategy by cacti. However, the flowers of cereoids are often large to huge in proportion to the stems

Figure 1.51. *Flower of* Cereus hildmannianus, *often called Peruvian Cereus. Flowers are followed by succulent fruit.*

Figure 1.52. *Flower of totem pole cactus* (Neobuxbaumia polylopha), *Barranca de Metztitlán, Hidalgo.*

Figure 1.53. *Flowers of xoconochtli,* Stenocereus stellatus, *Reyes Metzontla, Puebla.*

Figure 1.54. *Flower of* Stenocereus griseus, *Río Estorax, Querétaro.*

Figure 1.55. Roots of large sahueso (Pachycereus pringlei), El Cardonal, Sonora. Roots remain near the surface and may extend more than 10 m (33 ft.) from the trunk.

from which they emerge, so the brevity of their time on earth is somewhat compensated for by their majesty.

I believe that all columnar cacti produce edible fruits. Some fruits are grainy, dry, and tasteless; some are bland; and a few *(Corryocactus)* are decidedly sour. Most of the fruits, however, yield sweet, tasty pulp that varies in color from white to green to yellow to red to purple to sienna (tetechos). Most fruits also have husks that can be bristly (etchos), hairy (most viejitos), scaly *(jiotillas [Escontria chiotilla],* San Pedros, soberbios), spiny (most pitayas), woolly/spiny (sahuesos), or thick (most *Cereus* spp.). All these leathery covers demonstrate the plants' strategies for protecting the seeds and their matrix until the seeds are capable of germination. Many fruits split open when they are ripe, exposing the sweet pulp, which is often colorful and quickly attractive to birds and bats. In other cases, the hair, scales, spines, or wool soften and slough off, and the husk softens as well. The pulp itself ripens and changes color as the small to tiny seeds mature, meaning that fruit eaters (primarily birds, but also insects and several mammals—including humans) enjoying the succulent pulp will also ingest and partially digest the seeds. Digestion not only aids dispersal of the seeds (and assists recruitment), but also appears to help soften or dissolve the seed coat, promoting germination in some species. The scat in which the seeds

are defecated also may serve as a fertilizing medium. By eating the fruits in the wild, we are doing the plant a favor even as it bestows a favor upon us.

Roots of cacti in general and of columnars in particular deserve mention. Although the root structure of columnar cacti is not decidedly different from those of some other plant families, most of these cacti have very shallow roots, usually penetrating no more than a few centimeters below the surface. Indeed, the deepest recorded root for a cactus is 77 cm on a saguaro, about 2.5 ft.[49] Rather than penetrating to depths, most cactus roots fan out from the base of the plant and remain near the surface to capture rainwater. This shallow root system is quite effective, enabling the plants to take a drink after only minor rainfall, but it also makes them vulnerable to toppling over by winds. To compensate for this vulnerability, the large cacti send roots out horizontally, ranging as much as 15 m (50 ft.) and probably more from the base. Each of these roots (I have seen them at 10-m distance from the plant with a diameter greater than 8 cm [3 in.]) can act as an outrigger, stabilizing the plants during high winds. The critical time, though, is after substantial rain, when the cacti swell up with absorbed water. They assume enormous weights, and the wrong wind can topple the top-heavy creatures. (Fig. 1.55.)

Lightning also poses hazards for columnar cacti. Many, especially those of North America, grow in areas with intense summer thunderstorms and frequent lightning strikes. Ecologist Forrest Shreve believed that most saguaros die from mechanical injury, including lightning strikes.[50] Although shorter columnars may be exempted from lightning damage by their unobtrusive stature, and some giants grow where lightning strikes are rare or absent, tall columnars are often vulnerable. Old saguaros and sahuesos may be culled by being struck and ultimately killed by heavenly punishment. Even if only one saguaro from a grove is killed every few years, this number will amount to a high percentage over the lifetime of a cohort. The tallest columnars illustrated in this book, in figures 1.7 and 1.8, grow in canyon bottoms where they are less prone to be struck by lightning. The tallest *Pachycereus pringlei* reach 19 m (63 ft.) in Baja, California, where lightning strikes are rare.

Origin, Evolution, and Taxonomy of Cacti

Part of appreciating the natural history of cacti and especially of columnar cacti involves dealing with questions of the *origin* of cacti, the *mechanism* for their widespread distribution, and the *relationship* of species and genera to each other within the cactus family. These topics are interrelated, but I deal first with theories of origins and distribution of the cactus family.

For two centuries after Carl Linnaeus recognized cacti as a group in 1753, attempts to describe their relationships produced confusion about the family. Only in recent years has agreement about the phylogeny (evolutionary origins) of the cactus family based on genetic studies begun to emerge. Cactologists widely agree that cacti evolved from the portulaca family or a close relative in tropical or near-tropical conditions, probably in northwest South America or somewhat farther south in the central Andean or proto-Andean region.[51] The family appears to have evolved over a 40-million-year period in the late Cretaceous and early Tertiary 75 to 35 million years ago.[52] Other estimates place the family as old as 100 million years and as young as 20 million.[53] The strange, treelike *Calymmanthium* of the Río Marañon region in Peru or a similar extinct species may be the primitive ancestor of columnar cacti.[54] By the Pliocene into the Pleistocene (between 5 and 1 million years ago), columnar cacti appeared and filled ecological niches in the drying environments to the north and south. Intercontinental convergent evolution, not genetic closeness, resulted in the appearance of plants of remarkable similarity, such as the saguaro in the Sonoran Desert and the pasacana in the cooler high desert of northwest Argentina.[55] This result

Figure 1.56. Calymmanthium substerile *near Balsas, Río Marañon, Peru.*

means that species going their own separate ways and having no contact with each other evolved into plants with remarkable physical resemblance. (Fig. 1.56.)

The story of where, how, and to what extent the plants radiated from the South American locus of evolution, diverged from other cacti, and finally converged into new species quite parallel and similar to forms occurring on another continent is yet to be told. Re-creating the deep history of the cactus family is no simple matter because cacti are notoriously and frustratingly absent from the fossil record.

Why do cacti not show up as fossils when other plant families appear so widely and prolifically? First, they are largely succulent and leafless, and the process of fossilization favors plants with a greater proportion of woody structures and with leaves. Second, most cacti, especially the columnars, tend to grow outside of watercourses and wetlands. They prefer more xeric (dry) sites to the mesic (moist) conditions around the margins of inland seas, streams, swamps, and lakes, where the anaerobic conditions and the accumulation of sediments involved in fossil formation are more favorable. Cacti tend to love well-drained and dry places, and those sorts of locations are not ripe for fossil production. Even where columnar cacti grow in more tropical conditions, they tend to locate on rocky, well-drained hillsides where their green branches have better access to sunlight.

A third reason for the absence of cacti traces in the fossil record is that cacti have relatively large, soft pollen grains, which apparently are not easily amenable to fossilization in the way that small, tougher grains are. Tom Van Devender and others have documented fossilized pollen and seeds of cacti in packrat middens (packrat urine has remarkable preserving powers), but they date from no earlier than forty thousand years before the present and provide only limited information about the evolution of the plants involved. To date, no fossilized cactus pollens from rock samples have been identified.

With no fossils to help date the appearance of cacti in the evolution of vegetation, only less direct means are available to establish their phylogenetic origin. One such method of arranging related plants in chronological or evolutionary order is using comparative anatomical/physiological analysis to determine which plants and plant characters are more primitive (older) and which are more advanced or derived from more primitive ones, hence more recent on the evolutionary scale. Thus, several researchers consider the cactus genus *Pereskia* to be a primitive forbearer of cacti, for it has both leaves and a superior ovary, while nearly all other genera of cacti are leafless and have inferior ovaries. They trace *Pereskia*'s origin to northern South America. Considering its primitive structural features as the evolutionary ancestors of more specialized characteristics, they propose a radiation of cactus descendants of the primordial *Pereskia* throughout South America and into North America.

Geological tracing of present species' ancestors presents no mean problem. In the late Cretaceous (70 million years ago, the proposed approximate time of the origin of cacti), North and South America were separated by several hundred kilometers of ocean. Central America did not exist until the Pliocene, some 3–5 million years ago. How, then, did cacti arrive in what is now Mexico? One possibility is that plants, including cacti, bridged this water gap by "late Cretaceous island hopping along the proto-Antilles"; that is, the early cacti may have "hitchhiked" from South America to islands that existed then and thus reached the landmass of North America.[56]

Recent tectonic studies lend weight to this line of thought. The landmass that is now the Netherlands Antil-les and other Caribbean Islands is part of the Caribbean Plate.[57] It appears to have originated as the easternmost remnant of the Galápagos volcanic hot spot, now located a thousand miles to the southwest. In the late Cretaceous, around 90 million years ago, the massive block migrated eastward like an underwater barge, forcing the existing plate aside as an icebreaker crunches through ice, forging relentlessly through the gap now occupied by Panama and Central America and docking with the northern coast of South America. This navigating crustal mass, complete with island arcs (perhaps themselves large scars left by Galápagos vulcanism), would have provided an ideal jumping-off spot and perhaps the right climate for expanding cacti populations. The entire mass, including the Netherlands Antilles, was closer to the equator in the late Cretaceous, several hundred kilometers to the southwest of its current location, so in the cool Pacific it quite possibly presented a dry climate at that time as well as at the present.

The docking of the continental mass that was deformed and perhaps propelled by the ancient Galapagan hot spot may explain how columnar cacti arrived on the Caribbean islands, but if there had been no contiguous landmass for cacti to have colonized to the north, and other islands were more or less far removed from the South American continent, how might they have reached the new islands? In other words, how would the "island hopping" have occurred? One possibility is dispersal of seeds by birds or by floating logs or vegetative masses rafted from continental rivers. Naturalist Jonathan Weiner describes seeing from the air large masses of vegetation emerging from Ecuador's Río Guayas into the Pacific and being carried well toward the Galápagos Islands, 1000 km (620 mi.) west of continental South America, by the delta flows and ocean currents.[58] The even larger Orinoco River of Venezuela is of sufficient magnitude and power to raft large masses of vegetable matter well into the Caribbean, certainly as far north as Trinidad and Tobago, and quite probably well beyond. Birds, ocean currents, and trade winds undoubtedly contributed to the northward movement of the species as well (for example, the arrival of cacti in southern Florida). Leuenberger places the locus of origin of *Pereskia* in the western or northwestern Gondwanian continent (the supercontinent that formed during the Paleozoic era before South America separated from Africa and finally broke up for good toward the end of the Mesozoic era), probably in what would become northwestern South America.[59] Using this location for the origin of cacti or for a point in the expansion of cacti from a different geographical origin explains how the early cacti could have leaped to the proto-Caribbean islands, which at that time were probably either docked with

Figure 1.57. Barranca de Metztitlán, Hidalgo.

South America or situated very close to it, and eventually to North America (Mexico[60]). If cacti had originated to the east and south, they would have tended to migrate to Africa, which during the early evolution of the cactus family was situated far closer to South America than the latter was to the North American land mass. Africa has no native cacti (except for the enigmatic single species of arboreal cactus *Rhipsalis*), and any location of the progenitors of cacti farther south and east of the present Colombia and Venezuela would put the evolutionary locus too close to Africa to account for the absence of cacti there. Fortunately, present-day distribution of *Pereskia* fits in well with this hypothesis, and the Caribbean coastal areas of Colombia and Venezuela are rich in columnar cacti. Because other interchanges of plant and animal species between the two separated continents occurred in the same general area, this theory presents a plausible explanation, but one that requires further verification.

Tracking more recent cactus migrations in the Sonoran Desert region is simpler. The presence of cactus pollen grains in fossilized packrat middens (large permanent nests), although a method effective only where climates are sufficiently dry to preserve the middens, has shown that the distribution of cacti was considerably different forty thousand years ago. Noting where a species is now most commonly found and where it is rarest, and combining that information with semifossilized pollen data from packrat middens, one can infer the species' possible evolutionary origin. Similar studies might show how the rise of the Andes created deserts and semideserts, stimulating cactus species to proliferate throughout arid and semiarid South America, as well as how in North America multiple species of columnar cacti came to be associated in such Mexican evolutionary hot spots as the Cañon Infiernillo of Michoacán, Barranca de Metztitlán of Hidalgo, and the Valle de Tehuacán.[61] (Fig. 1.57.)

As for the origin of columnar cacti, some authors propose that the strange monotypic (one species) Peruvian genus *Calymmanthium* was the evolutionary progenitor of the giants. These spindly but rather tall plants are confined to a narrow elevational band in northern Peru. They may reach 8 m (27 ft.) in height and achieve treelike appearance, but the branches are narrow and have at most three or four ribs, not even vaguely reminiscent of columns. Current research gives strong backing to the hypothesis that it was from the gangly *Calymmanthium* that most, if not all, columnars evolved and began their march to warm (and some cool) portions of the Western Hemisphere. And if *Calymmanthium* is the ancestor of columnar cacti, the Peruvian Andes where *Calymmanthium* grows is the most plausible location for their origin.

That sums up the most widely accepted theories on the origin and distribution of the cactus family. More specific anatomical studies of cacti attempt to establish intercacti relationships—that is, to group genera and species of cacti to each other in evolutionary relationships, leaving to other researchers the questions of the biogeographical origins and the ancient national and international travels of cacti.

Botanists use flowers above all to distinguish among species. But cacti present such a large spectrum of flower variation within and among species that traditional sorting can be confusing. Flower studies have progressed from basic flower anatomy to microscopic analysis to genetic sequencing. All methods are still vital in determining relationship. Franz Buxbaum in the 1950s and 1960s and Arthur Gibson in the 1970s and 1980s delved deeply into the anatomy of cactus stems and flowers, using a microscope to study and compare seeds and flowers. They added biochemistry as a tool for determining the appropriate genus and species for a subject plant. Gibson went even further, incorporating the use of scanning X-ray photomicrography to study the seed coats of various species of the current genus *Stenocereus,* discovering that patterns of striations found on seeds were an apparently reliable guide to phylogenetic relationships.[62]

The upshot of these studies is that from the reproductive structure, stem anatomy, biochemicals, and seed coats, scientists can construct hypothetical phylogenetic "trees" (rather like genealogical trees) and demonstrate what evolved from what. Analysis of similarities and differences of cellular, floral, rib, seed, and spine structures among apparently related species can establish how closely (or distantly) the species are related, enabling the researcher to locate species' relative position in the evolutionary tree. The guiding hypothesis is that the fewer the characteristics a cactus has in common with the ancestral *Pereskia,* the more

recent is its evolution. For example, silica bodies are found in the stems of members of the genus *Stenocereus,* but are absent from other genera and subtribes, thus suggesting that phenomenon as a marker for the genus.

Using such a combination of gross anatomy, microscopic analysis, and photomicrography, Gibson proposed that columnars of Chile and Peru were ancestors to the Stenocerinae (including the genus *Stenocereus*) of Mexico and the Gymanthocereae (including the genus *Browningia*) of the Andes.[63] If Gibson is correct, a further problem arises: How can the species of *Stenocereus,* so far apart in their distributions from the species of *Browningia,* have been derived from an ancestor they share? Why is *Stenocereus* a dominant genus in North America and *Browningia* similarly important to the south? These questions refer us back to biogeography, ancient climates, and tectonic movements.

More recently, Wallace has joined others in using submicroscopic, or molecular, studies to establish phylogenetic lineages of cacti, as well as the relation of cacti to their evolutionary progenitors. His molecular studies of cactus DNA established the comparative presence or absence of specific specialized genetic sequences.[64] The cacti with the most primitive *(basal)* characteristics as revealed by genetic sequencing tend to be grouped in the central Andean region, which includes northern Chile, northwest Argentina, Bolivia, and Peru. If a preponderance of primitive cacti are concentrated in an area, there is reason to believe that more-developed species evolved from plants in that mother lode.

Using this information as an indicator of evolutionary lineage (or lack thereof), along with DNA sequencing, Wallace has constructed his version of hypothetical evolutionary trees. He hypothesizes that the cactus family evolved in the central Andes (Bolivia, Peru, and Chile) rather than in northwestern South America, and it radiated from that origin to the north, east, and south. Many, but by no means all, columnar cacti are derived from two distinct evolutionarily separate lineages. One such lineage gave rise to the tribe Pachycereeae, which includes most North American cereoid cacti. From the other evolved what Wallace calls the BCT Clade (the tribes Browningieae-Cereeae-Trichocereeae), which includes a majority of South American giants. Wallace's findings suggest that the Peruvian columnars of the genus *Corryocactus,* the sancayos of Peru, are apparently ancestors of the Pachycereeae, which is more ancient than the other North American lineages. *Armatocereus,* the sausage cactus of Ecuador and Peru, and *Jasminocereus* of the Galápagos Islands appear to be even older and are also ancestors to the Pachycereeae and

Figure 1.58. Melocactus salvadorensis, *Minas Gerais, Brazil.*

Figure 1.59. Sina in flower, Sonora.

to *Corryocactus* as well. *Stetsonia,* the toothpick cactus of Argentina and Paraguay, probably lies somewhere near the Cereeae on the family tree, but for the present sits off by itself. The large *Eulychnias* of Chile appear to be related either to Notocacteae, a more primitive tribe ancestor to the BCT Clade, or, more probably, rather distantly to the Pachycereeae, making for a large geographical disjunction. Genetic-sequencing studies appear to demonstrate conclusively that the similarities of pasacanas in Argentina and saguaros in Arizona are the result of convergent evolution and not evidence of close genetic relationship: they may look alike, but they are only distant relatives.[65] Molecular-based analysis also suggests that noncolumnar genera such as *Echinocereus* and *Melocactus* are derived from columnar ancestors, but had to settle for diminutive size: *Echinocereus* because it radiated into cold climates where columnars could not survive, *Melocactus* because it came to devote inordinate space to development of a proportionally huge cephalium. The mere fact that a plant is a columnar cactus is no guarantee that all its progeny are columnars or will continue to be. (Fig. 1.58.)

Another evolutionary approach to relationships among columnar cacti stresses environmental factors in the evolution of columnars. Certain columnar cacti (*Myrtillocactus geometrizans, Isolatocereus dumortieri,* and *Stenocereus queretaroensis*) have an affinity for noncalcareous or silicaceous (silica-rich as opposed to calcium-rich) soils. Researchers suggest that these species are phylogenetically less advanced

(i.e., more closely related to the ancestral *Pereskia*) than those that flourish on calcareous soils (*Cephalocereus columna-trajani, C. senilis, Mitrocereus fulviceps, Neobuxbaumia macrocephala, N. tetetzo,* and *Pachycereus hollianus*). The latter are "derived," or more recently evolved.[66]

Finally, each cactus has its own pollination strategy, which may hint at its peculiar evolution. Gibson and Nobel and others have used pollen characteristics (plus microscopic studies of seeds) along with gross flower structure to infer taxonomic relationships among the numerous species of *Stenocereus.*[67] For all their spectacular beauty, columnar cacti flowers exhibit remarkable diversity, even among species that are close evolutionary comrades. This variety is probably a function of coevolution with pollinators, the flowers evolving structures to attract, accommodate, and manipulate pollinators, and the pollinators evolving characteristics enabling them to exploit the available nectar resource.

For the most part, white flowers open at night and are pollinated by bats or moths or both, whereas colored (especially red tubular) flowers open during the day and are pollinated by hummingbirds or insects or both. Flowers with sweet scents attract moths and other insects, and those with less pleasant scents (to us) tend to attract bats.[68] In the Sonoran Desert, sahuesos are primarily bat pollinated; sinas with their red flowers are hummingbird pollinated; and pitaya agrias with white to pink flowers are moth pollinated. (Fig. 1.59.)

Columnars with deeply tinted flowers are uncommon and often of highly limited distribution (e.g., *Neobuxbaumia multiareolata* with its purplish blooms, *N. polylopha* with its reddish blooms, and *N. euphorbioides* with its purplish to reddish pink blooms). White flowers seem to predominate, and among Pachycereeae 70 percent of species are pollinated by bats (among other creatures). Bat pollination affords a notable advantage in that bats are far ranging, and the broad dispersal of pollen can be some assurance against incestuous pollinization and hybridization (plants may have incest taboos, too!). The nectar-feeding bats *(Leptonycteris curasaoe)* are terrific pollinators. If you are lucky enough to see them in action (with night-vision scopes), or if you see photos of them illuminated in action, you will often notice their dark heads covered with golden pollen. Because they are greedy sippers of nectar, to attract them the cactus must produce copious amounts of the sugary fluid, which for the plants is metabolically expensive. Nothing is free.

This account of pollinators and their clients is oversimplified, for a great variety of insects and other creatures visit all flowers and are potential pollinators. Yet each pollinator is attracted (or repelled) by the peculiar design and lifestyle of a particular cactus; its floral opening sequence; its floral aroma, color, and shape; arrangement of sexual parts; amount of pollen and nectar produced; length of the floral tube; abundance of flowers on the plant; and several other characteristics. Nature's ingenuity in designing new and different blooms to give a species a slight advantage in attracting pollinators is intriguing indeed.

A more specific account of how cacti great and small evolved from their ancestors and how they became distributed throughout the Western Hemisphere remains a fertile field for research. Techniques invoking (among others) tectonic history, anatomical and biochemical features, fossilized pollen grains, and ecological variables all combine to demystify the distribution of cacti in general and of the columnar cacti in particular. Scientists' unfolding discoveries will continue as they overcome the limitations in fossils established by nature.

How and why rapid speciation takes place in particular cauldrons of evolution—such as the Valles de Tehuacán and Cuicatlán in southern Mexico with at least eighteen species of columnar cacti,[69] the Río Marañon region of Peru with at least twelve, and the Sonoran Desert of the southwestern United States and northwestern Mexico with at least eight—makes for fascinating speculation and a gradually thickening ecological record. A few fossils would help immeasurably to clarify the emergence of the cacti in modern flora.

As more research into cactus ethnobotany is undertaken, I am convinced that evidence of anthropogenic (human) influence on the distribution of cacti will mount. I cite as evidence the semidomestication of cacti many centuries ago in southern Mexico, the close relationship between the Sonoran distribution of *Pachycereus pringlei* and *Stenocereus gummosus* and the historic range of the Seri people,[70] colossal proliferation of *Stenocereus griseus* in the Caribbean rim, and the close association of *Trichocereus atacamensis* with Dieguito *pucaras* in northwest Argentina and southwest Bolivia. *Cereus hildmannianus* is so widely planted for its fruits and its beauty that its origins are obscure. *Corryocactus brevistylus* of Peru and Bolivia appears to have benefited enormously from the terracing of myriad Andean slopes by Andean civilizations. The use of *Pachycereus marginatus* for fencing in central Mexico is so widespread that the origin of the cactus is difficult to ascertain. Finally, I suspect that the great cacti forests called *tetecheras* of the Valle de Tehuacán reflect pre-Columbian management practices of Popolocas, the same people who are probably a critically important variable in the expanding range of the *chende (Polaskia chende)*. But this sort of work on anthropogenics of distributions will have to await another day and researchers with funds to conduct more genetic studies. (Fig. 1.60.)

The fuzziness of cactus evolution has resulted in cactus taxonomy's being a historical mess. The taxonomic history of the great columnars (i.e., the way humans have classified them among plants) is intriguing largely because over the past century a great deal of political posturing has been employed as species have been kicked around among genera and subfamilies. Genera have come and gone. Species have been booted from one tribe or genus to another, and some have been abolished or assimilated to other species. Prolific research has come to naught as taxonomic theories have been undermined by the facts.

A classic example is the genus *Cephalocereus*. Its members are currently confined to Mexico (possibly extending into Guatemala) and consist of five species, all of which are treated in this volume. Throughout its rocky history, though, sixty-six species have at one time or another borne the label *Cephalocereus*. Most have been redirected to other genera, and perhaps a few have been shown to be confused with previously named species. At one point, one author claimed it had become a monotypic genus—in other words, that only one species remained in it.[71] This turbulent generic history exemplifies the fluidity of cactus nomenclature, which will one day undoubtedly affect this book also.

This history is worth reviewing because understanding cactus taxonomy is the basis for understanding how

Figure 1.60. Sancayos (Corryocactus brevistylus), *Cañon Colca, Peru.*

columnars fit in with other cacti and how cacti fit in with other plants. A meeting of worldwide cactus taxonomists in 1984 formed the International Cactaceae Systematics Group to address the confusion of cactus nomenclature and to produce a consensus on what cactus belonged to which genus, tribe, and subtribe. Edward Anderson, a member of this group, wrote the comprehensive book *The Cactus Family* (2001), reflecting this group's findings, much of which (but not all) I have accepted. Even with publications of the systematics group, however, widely differing labels are still applied to species, which sometimes makes communications confusing, just as the variety of cacti themselves makes things confusing.

The best-known early attempt to classify all cacti was made by N. L. Britton and J. N. Rose early in the twentieth century in their four-volume work *The Cactaceae* (1919–23) based on extensive explorations and collections in the New World. In developing their taxonomy, they lumped all columnar cacti into one genus, *Cereus,* thus positing a closer phylogenetic relationship among all columnar cacti than actually exists (note the use of the general term *cereoid* in reference to these cacti).

Later taxonomists with better access to modern travel, microscopy, biochemical analysis, and cladistics undertook the daunting task of classifying the cacti. One of them, the German Curt Backeberg, went on a rampage of splitting taxa of cacti into new groupings. In 1966, he proposed an ambitious new classification that created many new genera, much to the displeasure of more conservative taxonomists. Few of his new genera of columnars persist today. In the 1950s, the botanist Albert Friç published fifty-two new genera, only one of which is recognized today. The Mexican cactologist Helia Bravo-Hollis published the comprehensive three-volume *Las cactáceas de México* in 1937, with a second edition in 1978.[72] Her careful study did much to settle the confusion resulting from Backeberg's enthusiastic forays into new names. At the same time, the Austrian Franz Buxbaum attempted to classify Mexican columnars on the basis of evolutionarily related characteristics.[73] He also demonstrated important evolutionary differences between North American and South American columnars, thus classifying them in different tribes. Arthur Gibson and his associates accepted Buxbaum's methodology and refined it with more sophisticated techniques involving electron microscopy and biochemical studies.[74] As noted earlier, Robert Wallace used DNA sequences to establish phylogenetic relationships to group most North American columnars (but by no means all: *Pilosocereus* is excluded) into the tribe Pachycereeae and most South American columnars into the tribes Browningieae, Trichocereeae, and Cereeae.[75]

Field and laboratory work have combined to clarify some taxonomic confusion. Gibson and Horak and then Cornejo did intensive studies of the Pachycereeae (cereoids of greater Mexico and the Caribbean), establishing two major evolutionary subtribes, the Stenocereinae and the Pachycereinae as a basis for their classification.[76] All Pachycereeae (unlike the Cereeae and most Trichocereeae) have a ringlike skeleton made up of bundles of wooden rods, as can often be seen in dying saguaros that remain upright as their succulent flesh rots and falls away. Gibson and Horak also discovered that members of the genus *Stenocereus* could be characterized by the presence of silica bodies in the skin, a phenomenon found in no other cacti. Gibson and Nobel further proposed that the genus *Pachycereus* be defined as "a genus rich in alkaloids and having large, glossy, black seeds, stems that blacken rapidly when cut, and many shared characteristics of stem anatomy, areoles, and flower structure."[77] These characteristics, they believe, indicated specific paths of evolution.

Recently, based on the work of Gibson and Horak and others, the Cactaceae Consensus Working Group assigned the genera *Backebergia*, *Lophocereus*, and *Mitrocereus*, which have long been prominently monotypic genera (one species in each genus), to the genus *Pachycereus*, and they assigned much of the genus *Trichocereus* to *Echinopsis*, making the latter a huge, almost unwieldy taxonomic catchall. For the purposes of this book, I mostly retain the older genera.

The Future of the Great Cacti

What do native peoples think of the great cacti growing in their backyards? Because they have grown up in the shadows of the giants, they mostly take them for granted, as each of us takes for granted our native landscape. Attitudes vary, of course, but where climates are more or less arid and great cacti form an increasingly noticeable component of the vegetation, their human uses increase in importance.

In the Valle de Zapotitlán in Puebla, where no fewer than fifteen species of columnars can be found within a 16 km (10 mi.) radius, older natives are familiar with all the varieties and have common names for them. They were surprised to learn that outsiders viewed their landscapes with awe. In the 1990s, the Mexican government established a botanical garden nearby, set in a sensational forest of tetechos. The inhabitants view the institution with curiosity and a touch of bewilderment. Relatively few of them still gather cactus fruits, eat eggs scrambled with tetecho buds, or use tetecho wood in building houses, but all seem to respect the groves of cacti. Similarly, Argentines of Inca descent recognize the virtue of pasacana fruits, but for the most part they no longer collect them (one lamented to me, "No mas van a pasacanear") or harvest the hard and versatile pasacana wood, except for tourist crafts. Many Seris of northwest Mexico view with nostalgia the days only a decade ago when they ventured on foot into the monte to collect the succulent fruits of *ziix is ccapxl* (pitaya agria), *ool (pitaya dulce), mojepe* (saguaro), and *xaasj* (sahueso),[78] and when they carefully cleaned and preserved the seeds of thousands of sahueso fruits. They no longer practice collecting because they now have trucks to drive into towns where they can purchase food. Older Seris report that younger Seris disdain traditional cactus harvesting. Residents of an indigenous community in coastal Ecuador professed to me a fondness for the fruits of *Armatocereus cartwrightianus*, but acknowledged that they do not collect them as they did before market foods were easily available to them. Andean peasants in the Río Marañon region widely extol the virtue of purifying buckets of water by dropping in a section of *Armatocereus* branch, but very few continue to do so. They purchase bottled water instead. (Figs. 1.61 and 1.62.)

However, in Jalisco, Mexico, smallholders cultivate *Stenocereus queretaroensis* in densely planted orchards and harvest the pitayas commercially. Marketing these fruits is an important and expanding industry in the small town of Techaluta in the Valle de Sayula, where a pitaya cooperative has obtained machinery to process the fruits into preserves. In Oaxaca and Puebla, farmers are converting hundreds of acres of chronically overgrazed pasture to fields planted with at least four species of *Stenocereus* cultivated for their fruits, many of which are marketed commercially. The number of plots under cultivation is increasing as farmers realize that cactus fruits can bring a better monetary return than simple pasture or field crops. The Tohono O'odham of Arizona still gather saguaro fruits, and some individuals continue to participate in the annual wine festival. In parts of central and southern Mexico, wild babosos (*Pachycereus hollianus*), chichipes (*Polaskia*

Figure 1.61. Grove of tetechos (Neobuxbaumia tetetzo), *Zapotitlán de Salinas, Puebla, Mexico.*

Figure 1.62. Pitayas, fruits of Stenocereus thurberi, *Sirebampo, Sonora.*

Figure 1.63.
Neobuxbaumia
polylopha *raised from
seed, Bach's Nursery,
Tucson. To the right are*
Trichocereus terscheckii,
*often called South
American saguaro.*

chichipe), garumbullos (Myrtillocactus geometrizans), and pitayas are marketed as they have been for centuries. The most consistently collected is the fruit of garambullo, which is crushed and fermented into a popular wine with a taste rather reminiscent of cough medicine. The same is true of fruits of *Pachycereus weberi* and of some Chilean species of *Eulychnia.*[79] In general, however, rural folk gather fewer cactus fruits and products than they once did, which diminishes the social and economic significance of cacti, while increased population and grazing pressures imperil the survival of many plants. Perhaps only the lure of alcohol, stronger than even the pull of the computer screen, will be the salvation of the great cacti.

A different sort of threat to the great cacti has been poaching, or illegal collecting, of cacti. Cacti are wildly popular worldwide, especially in Europe (even more so in eastern Europe) and Japan. Clandestine trade in rustled plants, even mature columnar cacti, continues to pose serious problems for some species, in spite of the international adoption in 1973 of the Convention on International Trade

in Endangered Species of Wild Fauna and Flora (CITES), which has slowed but not eliminated the contraband traffic. In Brazil, Japanese collectors sent in large trucks to parts of northeastern Minas Gerais province and stripped these areas of great cactus diversity "without leaving even a seedling."[80] The government of Mexico has prohibited the export of any nondomesticated cactus, cactus parts, or seeds. Enforcement in Mexico and elsewhere, however, is spotty or nonexistent. Police bust cactus rustlers from time to time in my home state of Arizona, but they acknowledge that for each violator apprehended, ten go undetected.

The best remedy for poaching is the ready availablity of cactus plants legally raised from seed or cuttings. Columnar cacti are strikingly beautiful, require little attention except for protection from cold temperatures, attain great size, and live to ripe old ages. Their spines protect them from careless contact and herbivores, and their flowers are often a cause for celebration. The giants have a promising future in mountains, flats, parks, gardens, and patios. (Fig. 1.63.)

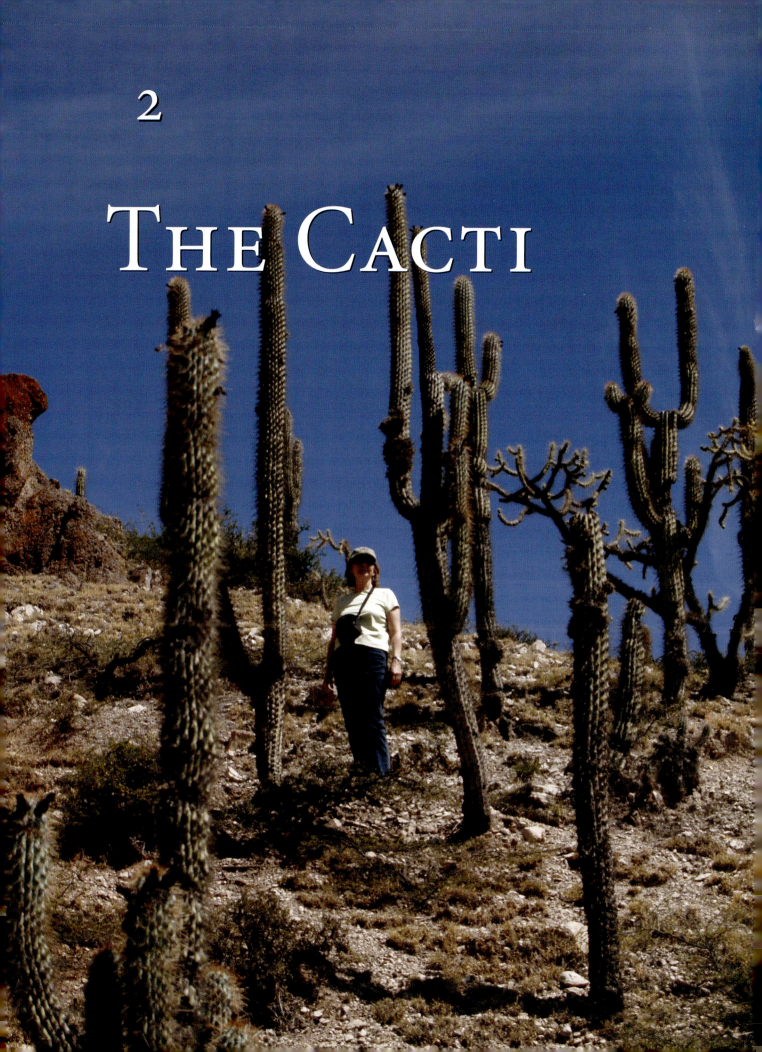

2

THE CACTI

ORGANIZING THE DESCRIPTIONS of more than one hundred species of cacti and how people use them has presented me with a number of options. The easiest way would simply be to list them alphabetically by genus and species. However, I wish to give special prominence to a small group of the most ethnobotanically important species and prefer not to be forced into alphabetical listing within genera. Presenting the genera by geographical region seems arbitrary. Although some genera tend to be geographically united, others cross regional boundaries; for example, *Pilosocereus* and *Stenocereus* are found in North America, South America, and the Caribbean. Presenting the cacti according to the tribes Cereeae, Browningieae, Trichocereeae, and Pachycereeae would present geographical anomalies and confusions. Because columnarhood is not a clear taxonomic concept, organization by taxonomy would also be difficult. Instead, I have chosen to present the plants by genera, beginning each with keystone species (cacti whose human uses stand out) and following with the other members of the genus. In some cases, I group genera by geographical region; in others, I present them in no particular order.

This treatment gives short shrift to South American genera. I have lived among and studied North American columnars for decades and have comparatively short exposure to the giants of South America. With a couple of notable exceptions, the ethnobotany of South American columnars is less varied. Why this is the case is difficult to explain. Except for the pasacana and perhaps the San Pedro cactus *(Echinopsis pachanoi),* no South American columnar cactus stands out as being culturally critical—that is, a plant so central to a culture that in its absence the culture would have developed differently or been forced into a different habitat.

The accounts here have glaring deficiencies. I have given inadequate attention to the columnar cacti of Bolivia, especially the charismatic *Neoraimondia herzogiana* and the monotypic genera *Samaipaticereus* and *Yungasocereus.* Travel logistics for studying the cacti in that country are complicated and discourage cactus exploration. Similarly, I present photos of fewer than half the columnar cacti of Brazil. Absent are the genera *Arrojadoa, Coleocephalocereus, Espostoopsis,* and *Siccobaccatus.* Although such small columnars are numerous in that vast nation, they are for the most part widely dispersed and restricted in their habitats, so that photographing them would represent a short lifetime's work. Furthermore, they tend to be isolated populations and appear to have few or no human uses. (Fig. 2.1.)

Nor have I done justice to the cosmopolitan genus *Cereus.* It, even more than the *Pilosocereus* and *Facheiroa* of

Figure 2.1. Neoraimondia herzogiana, *Bolivia's tallest cactus.*

Brazil, is widely dispersed and, apart from a few well-known species, of minor significance ethnobotanically.

The index is indispensible in locating on which pages each species is treated. Common names, vital to ethnobotanical studies, are listed in the glossary. Ethnobotanists typically divide uses of plants into categories such as food, medicine, religion, culture, industry, construction, and so on. However, in only a few cases have I been able to identify multiple nonfood uses for columnar cacti. For the most part, when cacti are used, it is primarily as a food source. Consequently, I do not follow the plant's description by a listing of use categories. I simply describe each species, its distribution, and the ways in which I have found native peoples using it. For some species, I could discover no uses. I nevertheless include descriptions of these plants, for I am convinced that everyday uses (as opposed to biochemical or other such commercial uses) will someday be recognized, at least for those species that survive the onslaught of habitat destruction savaging the planet.

To avoid overpowering the text with equivalent measurements, I have given only the metric measurements in this chapter, but a general guide to follow is: 250 mm = 10 in., 20 cm = 8 in., 20 m = 66 ft., 1500 m = 5000 ft., and 100 km = 62 mi.

Saguaros

Carnegiea gigantea

saguaro, *hashan* (O'odham), *mojepe* (Seri), *saguo* (Cáhita)

I begin with the saguaro for both chauvinistic and practical reasons. It is the world's best-known columnar cactus, primarily because it is the only common columnar cactus found in the United States. It is also the most studied columnar cactus species, especially the populations in southern Arizona. Historian Bernard Fontana set out to compile a bibliography of saguaro ethnobotany and found more than two hundred publications.[1]

Although the saguaro is found only in Arizona and Sonora, Mexico (with a few individuals in California), it is widely represented as ubiquitous in the U.S. West in popular depictions of the desert vegetation and in movie "Westerns." The California population is limited to a couple of locations a few miles west of the Colorado River. The largest population, across from the mouth of the Bill Williams River in northwestern Arizona, appears to have originated from seed brought downstream on floods through Arizona.[2]

Due to their large size (unusually tall saguaros are among the tallest cacti) and peculiar ecological requirements, mature saguaros usually do not respond well to transplanting or cultivating outside the greater Sonoran Desert. The oddness and massive size—the mystique—of the cactus have enhanced its commercial value, and land developers and landscapers have attempted to transplant many thousands of mature plants, with a rather low rate of success for individuals with branches. Because the large plants are slow to die—they may languish for several years before offering their souls to the heavens, turning brown and shedding their flesh or falling over—transplanters may claim success where their efforts are realistically failures. Transplants' failure to survive and reproduce is just as true outside the saguaro's range in semitropical Mexico, where residents often attempt to grow them as ornamentals, as in the United States, where having a saguaro cactus in one's yard is often a status marker of sorts.

Saguaros grow abundantly in the lower Sonoran Desert, ranging from northwestern Arizona south to southern Sonora, a distance of 1000 km. They grow well in bottomlands and on gentle slopes in the middle of their range (primarily in western Sonora) and on gradual to steep slopes at the northern and southern extremes. In the driest, hottest parts of the Sonoran Desert, they survive only along the margins of ephemeral desert washes. The cacti flourish over a range in elevation from the shores of the Sea of Cortés to more than 1500 m elevation on protected southern exposures in the Hualapai Mountains of northwestern Arizona and similarly protected exposures in the Santa Catalina and Rincon mountains of southeastern Arizona. (Fig. 2.2.)

Some of the largest saguaros are found near the northeastern limits of the plant's distribution at the point where freezes preclude the survival of the frost-sensitive plants. An individual said to be more than 50 ft. tall fell over in the early 1980s near Cave Creek, Arizona, north of Phoenix. Another renowned individual in the same region, perhaps the tallest cactus of all, toppled over and was then measured at 68 ft. A 52-foot-tall saguaro with fifty-two arms, well known in the 1950s, grew in Saguaro National Monument in the Rincon Mountains east of Tucson near the eastern limit of the plant's range.[3] A celebrated plant more than 45 ft. tall with more than fifty arms with whom I had exchanged pleasantries over the years died in the late 1990s. It grew at 1200 m in Saguaro National Park and was known as "The Granddaddy."

Although these tall individual saguaros are not uncommon, the vast majority of plants are less than 9 m tall. Young saguaros grow as a straight (sometime club-shaped) stalk or trunk. When they are forty to seventy-five years old (or even older) and about 2 m tall, they usually begin reproductive growth and may sprout arms. Individuals under cultivation with supplemental watering may flower at twenty years of age. They continue to add arms throughout their life, usually a maximum of around two hundred years. Populations on flatlands in more arid areas, however, seem reluctant to grow arms, as is the case with the healthy but aged population growing at 400 m elevation in the arid flat valley 20 km west of Sonoyta, Sonora, just south of Organ Pipe Cactus National Monument in Arizona. The tall plants (up to 10 m) there develop few, if any, arms, and these arms tend to be located toward the *top* of the plant. Some arms may exceed the height of the central trunk, a habit unusual elsewhere. This pattern gives the plants a thinner, less massive, and more fantastical aspect than the traditional candelabra appearance. I found that in the vicinity of the very arid (75–100 mm annual rainfall) Tinajas Altas range of southwestern Arizona, only around 10 percent of saguaros have arms. In some areas

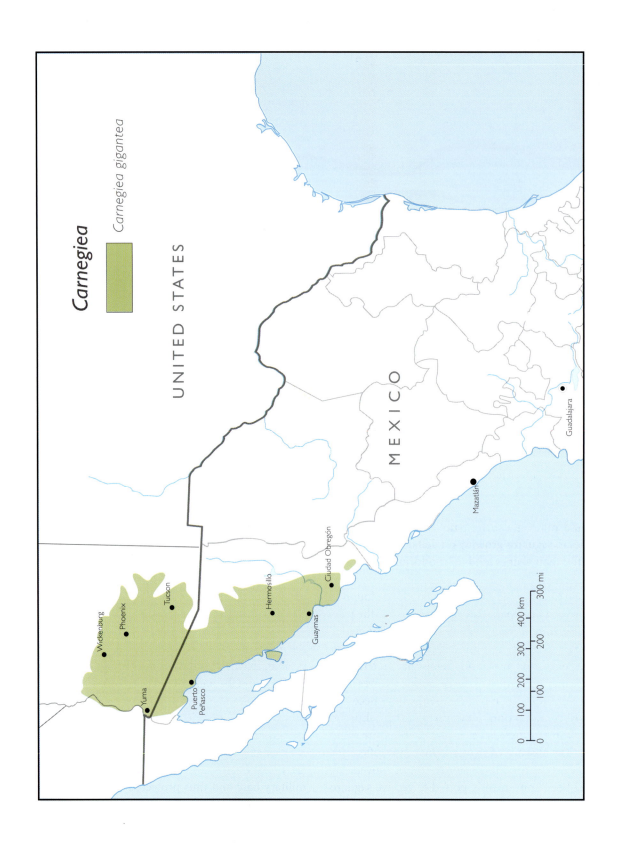

Carnegiea

Carnegiea gigantea

UNITED STATES

MEXICO

Guadalajara

Mazatlán

Ciudad Obregón

Hermosillo

Guaymas

Tucson

Wickenburg

Phoenix

Yuma

Puerto
Peñasco

0   100   200   300   400 km

0   100   200   300 mi

Figure 2.2. Unusual growth forms of saguaros (Carnegiea gigantea), *El Cardonal, Sonora. Low and erratic precipitation often produces such unusual appearances.*

there, I could find only single-trunked individuals. The same applies to saguaros growing on deltaic soils near the Sea of Cortés: their shapes tend to be odd, and their arms are few or absent.

I have long found it odd that the saguaro is not found in Baja California. The apparent explanation is that the peninsula's erratic rainfall may produce droughts that exceed the saguaro's tolerance, and annual rainfall throughout most of the peninsula's lowlands is below that preferred by saguaros. Furthermore, most of the peninsula receives little, if any, summer monsoon moisture, an apparent necessity for saguaros' reproduction. However, pitayos, sahuesos, and sinitas flourish there, so it is ideal habitat for other columnars. The absence of saguaros may also reflect the peninsula's deep history. In its northwestern movement, it has never been conveniently poised to receive saguaro seed in numbers that would establish a viable (genetically tolerant) population.

Saguaro National Monument was established in 1933 to protect a magnificent forest of saguaros on the western slopes of the Rincon Mountains, some 25 km east of Tucson. Nearly all these giants have died off and are remembered only in photographs, and the monument, now the east unit of Saguaro National Park, is nearly devoid of giant individuals. One of the densest populations of saguaros is found in the west unit of Saguaro National Park, some 50 km away. There and in adjacent Tucson Mountain Park the plants are especially abundant on volcanic slopes and bajadas. Extensive stands of flourishing, large individuals can be found throughout northwestern Sonora east of the Pinacate Volcanic Range. Thick growths are also to be found on steep hillsides of Pleistocene sediments in the Altar Valley of northwestern Sonora in Mexico. Exceptional densities are found in the Sonoran Desert National Monument near Casa Grande, Arizona, in an area that for decades was a military base and thus protected from livestock grazing and commercial development. Another vigorously healthy and dense population mixed with occasional organ pipes is found on a variety of hilly substrates just east of Magdalena, Sonora, continuing southeast 50 km to the Río San Miguel.

When the four-lane International Highway was constructed in the early 1990s, it bisected the marvelous stand of cacti. Along the highway, the great cacti can plausibly be said to grow in forests.

In southern Sonora at the southern limit of their distribution, saguaros grow only on basaltic slopes and outcrops where high temperatures and thin soils give them a competitive advantage over otherwise larger trees and shrubs of the thornscrub. The plants grow scattered, clearly visible from a distance in these habitats, especially during the dry season, when their bright, light green color stands out in contrast to the reddish brown of the basalts, the gray-brown of the dormant thornscrub, and the prevalent purple of the very common small tree *Bursera laxiflora,* usually leafless during drought. Natives consume saguaro fruits in diminishing numbers as one approaches the southern end of its range, not so much because of scarcity but due (they say) to the superiority of the fruits of the far more common pitaya and etcho.

Very young saguaros, like most columnar cacti, are especially susceptible to trampling and removal of protective cover ("nurse" plants) by livestock. As a result, recruitment of new plants over much of their range was severely hampered throughout much of the twentieth century. The healthy mix of ages—including juveniles—growing on Mesa Masiaca in southern Sonora, the southernmost population, appears attributable to the relative absence of tall and spreading tropical trees as well as to the fact that the rocky slopes are inaccessible to cattle and the trampling action of hooves.

During the middle of the twentieth century, biologists became alarmed at the sharp decrease in numbers of plants in Saguaro National Park. Various theories were put forth to explain the catastrophic decline from thick groves of massive plants in the 1930s to only a fraction of their numbers in the 1960s and a notable absence of juveniles. Indeed, visitors would arrive at the park headquarters in the 1970s to find only an occasional giant cactus. Since cattle grazing was eliminated in the 1970s, recruitment has improved, and by the end of the twenty-first century the park may once again see fine stands of large, many-branched giants. Unfortunately, introduced Mediterranean grasses, including red brome *(Bromus rubens),* have found saguaro habitat much to their liking since the late 1960s. Such grasses proliferate in the Mediterranean-type climate of winter rains, turning the landscape a dark green in often pure stands. During the spring drought, the grasses die back and become highly inflammable. The resulting fires—they spread quickly and burn hot—do not harm the grasses but usually prove lethal to saguaros and other

desert growth that evolved in the absence of fire. Fueled by foreign grasses after a wet winter, a massive range fire in 2005 charred tens of thousands of acres of fine Sonoran Desert north of Phoenix, killing thousands of saguaros in the process.

Studies in the vicinity of Tucson revealed that saguaro seeds successfully germinate only under ideal climatic conditions.[4] Seedlings appear to be established only when summer rains are sufficient both to germinate the seeds in late June and July and to provide ample surrounding plant growth to protect the tender seedlings and young plants. In the absence of these "nurses," the seedlings may become desiccated and die or may be nibbled to the ground by rodents and ants. Optimum conditions for establishment of surviving plants may occur but a few times each century. Whether these narrow requirements for recruitment of seedlings apply to populations growing at the middle and southern margins of the Sonoran Desert has not been determined. The thriving population of mixed ages (but apparently with no cohort of very young plants) near Magdalena at 650 m in northern Sonora endures intense livestock grazing. (Fig. 2.3.)

Drought increases saguaros' vulnerability to disease and death, and the plants' survival may be jeopardized by excessively high temperatures, but the most important limiting factor at the northern and northeastern limits of the saguaro's range is freezing and below-freezing temperatures. Although cacti in the northern portions of the Sonoran Desert endure frequent frosts, extended periods of below-freezing temperatures will kill the plants. In these extreme situations, healthy plants are often found growing on hillsides well above watercourses, but none is to be seen near the low-lying areas where cold air flows down from higher mountains upstream. Exposure of saguaros for one hour to a temperature of -9°C (15°F) results in immediate death of the cells and usually kills the plant.[5] A study of saguaro mortality following a catastrophic freeze in 1971 (temperatures fell as low as 11°F in some saguaro habitat) revealed that more than 50 percent of thirty-month-old naturally germinated plants in a study plot died as a result of the freeze.[6] Seedlings, very young juveniles, and very old adult plants are most affected by freezing, but middle-age plants are relatively immune from deleterious effects of normal minimum desert temperatures. At the other end of the thermometer, saguaros can tolerate temperatures as high as 65°C (147°F) for one hour before cells begin to die rapidly. The limiting factor of westward/northwestward colonization of saguaros in the very hot desert of southwestern Arizona and northwestern Mexico appears to be lack of rainfall, not excessively high temperatures.

Figure 2.3. Healthy mix of young and older saguaros. Tucson Mountain Park, Arizona.

Populations of saguaros I examined in the hot and arid Tinajas Altas Mountains of extreme southwestern Arizona showed widespread damage at their bases during the profound drought that affected the area between spring 1993 and 2003. Rodents and even desert bighorn sheep appear to have attacked the bases, sometimes gnawing away all the flesh and leaving a nearly bare space occupied only by the exposed ribs. How this affliction will affect cactus mortality will not become evident for several years due to the long period of senescence in saguaros. Saguaros appear to be capable of recovering from severe damage by producing large amounts of scar tissue, but it remains to be seen just how much mechanical damage they can tolerate.

At the southern end of saguaros' range, the factors limiting further southern expansion appear to be rainfall and other tropical plants with which the slow-growing saguaro cannot compete. The more abundant summer rainfall may also tax the plants' water-intake capacity beyond their ability to sustain the weight of the absorbed water. At Tucson, toward the eastern limit of the saguaro's range with 300 mm annual rainfall, 58 percent of precipitation occurs in the summer months.[7] At Masiaca, near the saguaros' southern limit with slightly greater rainfall (350 mm), nearly 80 percent of precipitation is concentrated in the summer.[8] In this region, the summer rains rapidly transform a withered, sun-blasted landscape of nearly unending grays and browns into a myriad of greens.

Lightning is probably a major cause of saguaro death through the plant's range, but especially in areas where they tend to grow tall. Although only a tiny proportion of saguaros are struck each year, the great cacti's longevity greatly increases the odds that over a lifetime they will be struck. Studies have revealed that lightning tends to weed out older individuals.[9] A strike may leave the plant intact, but weaken it so that its death later may be attributed to old age or bacterial infections.

The saguaro was originally classified as *Cereus giganteus* by the nineteenth-century botanist George Engelmann. In the 1920s, however, N. L. Britton and J. N. Rose, New York botanists who wrote the then-definitive work on the Cactaceae, changed the name to *Carnegiea gigantea*. The move was a crass example of political taxonomy. The new generic name was bestowed on the cactus with the hope that Andrew Carnegie, who was soon to visit the Desert Laboratory in Tucson (which he had funded generously), would be moved to continue or augment his largesse to the laboratory. It did not take Carnegie long to discover that the recent name change had little to do with botanical insight and lots to do with hopes for funding. Instead of being impressed, he was indignant at the shameless pandering to his presence in the desert. In spite of its shady origins, the name has stuck.[10]

Taxonomists now find strong evolutionary connections between saguaros and the genus *Neobuxbaumia* of southern Mexico, locating the saguaro in the same evolutionary branch as *Neobuxbaumia squamulosa*. The closest plants of this species grow along the coast of Colima and Michoacán, a disjunction of about 1100 km. I suspect that some day soon a bold taxonomist will pluck up her or his courage and proclaim the genus *Carnegiea* a failed political experiment and pronounce the saguaro to be *Neobuxbaumia gigantea*, even though *Carnegiea* is the older generic name. Buxbaum was a better cactologist than Carnegie. Taxonomists Barthlott and Hunt suggest that *Neobuxbaumia* is "perhaps better combined with *Carnegiea*."[11]

Except for occasional unconventional individual plants, saguaros flower in mid- to late spring, especially in May. Their white flowers open at night and are visited by a variety of pollinators, the most important of which are nectar-eating bats *(Leptonycteris curasaoe)* and moths. The fruits ripen in late June and early July and are eaten by the regular Noachian group of animals. The most noticeable are coyotes and foxes of the desert, for which the fruits become the principal dietary component, and the Gila Woodpeckers and White-winged Doves that seem to occupy every fruiting branch.

The flowers and fruits are concentrated near the tips of the branches. Each branch may house fifty buds or more, and the flowers on each branch usually open within a few days of one other, several at a time. At blossoming time, the desert landscape becomes suddenly studded with innumerable white patches of radiant flowers. When the fruits ripen and the husks split open, that same landscape becomes tinged with a most agreeable red. With the simultaneous ripening of millions of fruits, many remain unharvested by humans and animals and become desiccated while still on

Figure 2.4. Saguaro flowers. They usually appear in April and May.

the tree. Desert peoples (O'odham and Seri) harvest and save these naturally dehydrated fruits. The fruits were (and to a small extent still are) gathered by the Tohono O'odham and in addition to being eaten were processed to encourage fermentation into a wine of low alcoholic content. In mid-November 2000, while visiting the Cabeza Prieta National Wildlife Refuge in southwest Arizona, I found a saguaro with a large ring of ripe and past-ripe fruits circling the crown. The fruits, though bright red, were odd, tasteless, and riddled with holes. (Fig. 2.4.)

Much has been written about the role played by saguaros among native peoples.[12] In June 1970, I trucked a group of Seris well into the desert near the village of El Desemboque on the Gulf of California on a saguaro-fruit-gathering expedition. It was a joyous occasion for a very happy people; the saguaro fruits ripen after those of the sahueso *(Pachycereus pringlei)* and before those of the pitaya *(Stenocereus thurberi)*, thus providing fruit when there is little else available in the desert to satisfy the Seris' formidable sweet teeth. They consumed large numbers of fruits on the spot and brought many more back to their village for later consumption and drying. Their consumption was not associated with any particular ritual, but considerable interest was focused on my Land Rover that bore the harvesters home! (Fig. 2.5a.)

Although many other indigenous groups (Apaches, Hualapais, Maricopas, Pimas, Yaquis, and Colorado River peoples) regularly harvested the bright reddish fruits, none found the plants to be so central to their way of life as the

Figure 2.5a. Saguaro fruit. Note the dried pulp still within the split rind.

Figure 2.5b. Saguaro fruit.

Tohono O'odham (Papago) of southwestern Arizona and adjacent northwestern Sonora.

The Tohono O'odham were semiagriculturalist Uto-Aztecan speakers, along with their close relatives the Akimel O'odham (Pima), Hia-Ced O'odham (Sand Papago), and Ob (Mountain Pimas of Mexico). Prior to the Spanish Conquest, they occupied what are now the deserts and uplands of south-central and west-central Arizona and north-central and northwestern Sonora. The Pimas carried on intensive farming along the river systems of what is now central Arizona, and the Sand Papagos survived as hunters and gatherers in the dry deserts of what is now western Arizona and extreme northwestern Sonora.

The Tohono O'odham harvested extensive crops of tepary beans, squashes, and cotton. Their lands for the most part lacked permanent streams, but the O'odham selected crop varieties that would grow in the rich, moist deposits left by desert flash floods. They were heavily dependent on abundant summer rains and came to associate the ripening saguaro fruits with the onset of the rainy season. In an annual ceremony of great importance, the fruits were gathered in immense numbers and pressed. The expressed liquid was allowed to ferment and was drunk in the belief that the celebration would encourage the onset of summer rains:

> The fruit from which the ceremonial wine is made comes from the giant Saguaro cactus, a plant that is not cultivated but rather grows wild in the desert. The appearance of fruit on this cactus comes during a period of food scarcity due to the absence of rain. The fruit was processed into syrup, jam, dehydrated pulp, chicken feed, seed flour, oil, pinole, atol, snack foods, soft drinks, wine and vinegar, only the outer husk of the fruit going to waste. And even in discarding the fruit husk, a good result is thought to inure to these agricultural people, as it is always laid on the ground with the red inner part pointing upward "to hasten the rains."[13]

The saguaro was of such monumental importance to the Tohono O'odham that their annual calendar was based on the growing/producing cycle of the saguaro rather than on the progress of the sun. The old year ended and the new year began with the beginning of the saguaro fruit harvest. Many traditionalists believe that after death people become saguaros.[14]

Desert peoples frequently dried or preserved saguaro fruits for later consumption. For the most part, the Tohono O'odham who continue the fruit-gathering tradition harvest the fruits, which ripen at the tips of the branches, with poles formed by ribs of dead saguaros lashed together. A small crosspiece is lashed to the end, which enables the harvester to wiggle and push the fruits, while a child stands below with a basket or bucket to catch the fruits and keep the pulp from becoming soiled. Some indigenous peoples pressed the fruit and dried it, forming the seedy pulp into cakes. Others stored thickened syrup in sealed ollas (pottery jugs). Hodgson reports that the Tohono O'odham ground saguaro seeds and mixed them with other grains as a basic breakfast food.[15] O'odham knowledge of critical cooking and drying times during processing of the fruits enabled them to store large quantities of seed, syrup, and dried fruits without losing flavor and nutrition. (Fig. 2.5b.)

Without the fruit of saguaros, Sonoran Desert cultures could not have developed as they did or sustained their level of population. Bernard Fontana made the following report:

> To say cacti were important to our predecessors would be a great understatement. Among Tohono O'odham, harvesting saguaro fruit was much more than a prop for Arizona Highways articles and essays by ethnobotanists. In 1929, Thackery and Leding estimated that some 600 Papago families harvested saguaro fruit, collecting more than 100,000 pounds of the stuff each year. Castetter and Bell upped their estimate to 600,000 pounds annually. That's SERIOUS business.
>
> I once had a good helping of saguaro fruit in December, by the way. It was in Topawa in Laura Kermen's kitchen where she kept a few trays of frozen saguaro fruit cubes (instead of ice cubes) in the freezing compartment of her refrigerator. Best damned popsicles I ever ate in my life.[16]

In addition to the edible fruits, the "ribs" or rigid poles that remain when the plant dies were an essential building material throughout the drier parts of the Sonoran Desert. Thousands of ramadas (roofed, open structures built for shade) and buildings used the ribs as *latas,* or crosspieces, laid on top of mesquite beams to form the basis of the roofs. The ribs were also woven or stacked to provide walls. They are still used today by O'odham and Seris and by neo-Europeans seeking to create an appearance of authenticity in desert buildings.

Saguaro National Monument (it became a national park in 1994) was established by presidential decree in 1933. It and its sister Organ Pipe Cactus National Monument only a hundred miles away were the only national parks in the world dedicated to cacti until Los Cardones National Park was established in Argentina in 1994 to protect and enshrine the pasacana *(Trichocereus atacamensis).*

The origin of the term *saguaro* is uncertain. The plants were referred to by early Spaniards as *pitahayas,* a name of West Indian origin applied to many columnars of Mexico, mostly of the genus *Stenocereus.* Father Eusebio Kino referred to the plant as *pitahaya,* so that name was in use at the beginning of the eighteenth century. The Cáhita (Mayo and Yaqui) term for *Carnegiea* is *saguo.*

Pitayos and Pitayas: The Genus *Stenocereus*

This wide-ranging genus makes a considerable contribution to the flora of greater Mexico. Members of the genus are the best cactus fruit producers and are the Mexican columnar cacti most likely to have been domesticated. They range from the Caribbean Islands and Caribbean Rim (i.e., coastal Venezuela and Colombia and inland dry valleys) through most of Mexico into Baja California and extreme southwestern Arizona. (All columnar cacti are absent from Mexico's northeast.) I consider nineteen of the more than twenty-four species. As I note later, some taxonomic confusion remains within the genus, especially among *S. eichlamii, S. laevigatus, S. pruinosus,* and several species found in the valleys and lowlands of Oaxaca and Chiapas, none of which conform neatly to currently described taxa.

Many of the species (and most of the fruits) of *Stenocereus* bear the common name *pitayo* (the plant) or *pitaya* (the plant and the fruit) or *pitahaya.* In Mexico, the term *pitahaya* is used to refer to a vinelike cactus that produces enormous fruits *(Hylocereus undatus),* whereas *pitaya* refers primarily to *Stenocereus* fruit. The former term is perhaps derived from Arawak or Taino or another Caribbean language because the first reference appears from Puerto Rico in 1582.[17]

Pitaya Dulce: The Organ Pipe Cactus of Northwest Mexico

Stenocereus thurberi (Engelm.) Buxb.
organ pipe cactus, pitaya, pitayo, pitaya dulce, *aaqui* (Mayo, Yaqui), ool (Seri), tcutcuis (Tohono O'odham)

The organ pipe cactus is the most abundant columnar cactus in mainland northwest Mexico. It ranges from southwestern Arizona (where it has a national monument named after it) as far south as central Sinaloa and the drier portions of the canyon country of southwest Chihuahua. It is also common in much of the southern half of the Baja California peninsula. The largest populations (and the largest plants) are found in Sonora. (Figs. 2.6 and 2.7.)

The plants occupy a large range of habitats: organ pipes grow interspersed with pine and palms at 1100 m elevation in the foothills of the Sierra Madre Occidental. They survive on outcroppings in steamy tropical deciduous forest and survive in the extremely dry desert of the Pinacate Volcanic Range. Millions grow on the lowlands adjacent to the Sea of Cortés. They persevere in increasingly dense thornscrub into central Sinaloa, where they vanish in a scalped landscape from which all native vegetation has been removed. Where hills commence to the south and natural landscape can once again be seen, they are absent. (Fig. 2.8.)

The organ pipe varies from a multistemmed shrub branching from the ground where frosts are likely and in

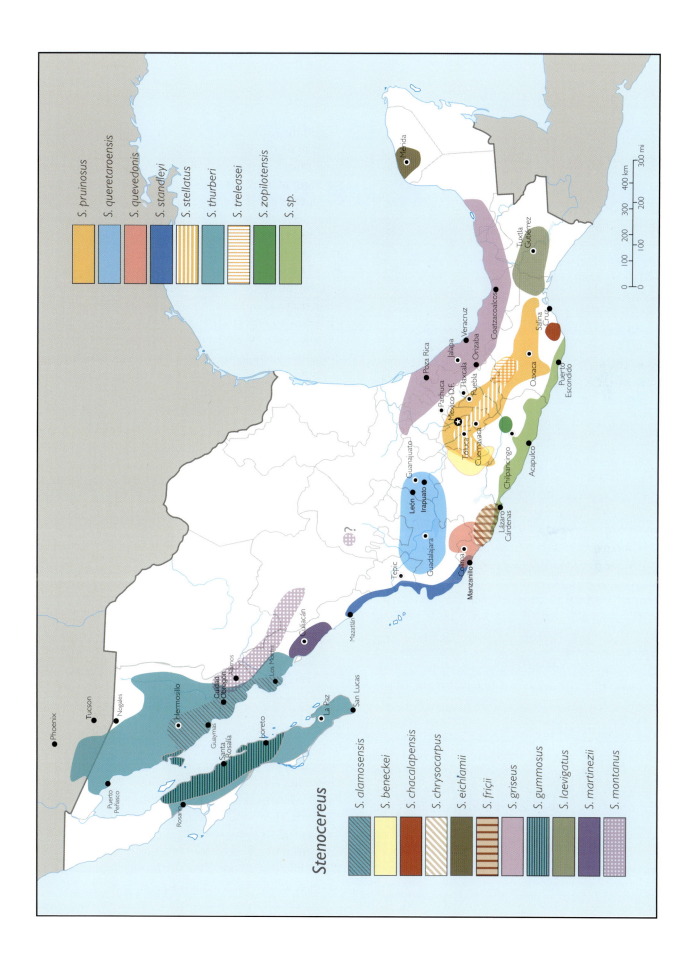

S. pruinosus
S. queretaroensis
S. quevedonis
S. standleyi
S. stellatus
S. thurberi
S. treleasei
S. zopilotensis
S. sp.

Stenocereus

S. alamosensis
S. beneckei
S. chacalapensis
S. chrysocarpus
S. eichlamii
S. friçii
S. griseus
S. gummosus
S. laevigatus
S. martinezii
S. montanus

Phoenix
Tucson
Nogales
Hermosillo
Guaymas
Santa Rosalía
Loreto
Puerto Peñasco
Rosario
La Paz
San Lucas
Ciudad Obregón
Alamos
Los Mochis
Culiacán
Mazatlán
Tepic
Guadalajara
Colima
Manzanillo
León
Irapuato
Guanajuato
México D.F.
Pachuca
Toluca
Cuernavaca
Chilpancingo
Lázaro Cárdenas
Acapulco
Puerto Escondido
Oaxaca
Puebla
Tlaxcala
Orizaba
Jalapa
Veracruz
Poza Rica
Coatzacoalcos
Salina Cruz
Tuxtla Gutiérrez
Mérida

?

300 mi
400 km
300
200
100
200
100
0
0

Figure 2.6. Large organ pipe (Stenocereus thurberi), *southern Sonora. Natives report that even larger plants are found in a remote canyon.*

Figure 2.7. Organ pipe cactus forest, known locally as El Pitayal, Masiaca Indigenous Community, Sonora.

Figure 2.8. Organ pipe cactus, Himalaya, Sea of Cortés, Sonora.

the drier portions of Sonora (less than 150 mm rainfall) and Baja California to an 8–10 m tree in coastal thornscrub. In even wetter foothills thornscrub and tropical deciduous forest, it is less common, but grows to very large size, as tall as 12 m in height, some individuals developing hundreds of arms. A subspecies *(littoralis)* of the plant found on cliffs along the extreme southern Pacific coast of Baja California is much reduced in size, reaching a maximum of a couple of meters in height. (Fig. 2.9.)

Larger specimens, especially those growing in the uplands of southern Sonora where frosts are rare, sometimes possess a thick trunk rising up to a meter above the ground. In general, though, the arms of pitayas branch closer to the ground, than do those of other columnar cacti that share its habitat, which invariably have a discernible trunk. The pitaya branches also curve and tend to be thinner and diverge from the vertical axis at a greater angle than etcho and sahuira branches, which tend to emerge horizontally from the bole for a short distance and then grow upward parallel or at an acute angle to the main axis. At times, organ pipe arms grow in many different directions. Their numerous ribs (twelve to nineteen) afford a smoother appearance to the branches when seen from a distance than is the case with other columnar cacti in the region, except for the saguaro.

Many large organ pipes are more than one hundred years old. Mayos who have lived in the same locations for generations and know their neighborhood plants intimately can verify such ages. The plants appear to grow quite slowly in the first decade, then move into a period of faster growth. By the end of the first seven to eight years, most plants are 30–40 cm tall, more than one foot. Growth can then become rapid, especially in years with average or above-average rainfall. Enriquena Bustamante, a Mexican scientist, has determined that pitayas show average growth rates of .03 cm per day near Organ Pipe Cactus National Monument, and .12 cm per day in the Pitayal, the organ pipe forest of southern Sonora.[18] In other words, the southern plants grow four times as fast as the northern ones, averaging about 40 cm a year in the Pitayal and about 10 cm a year at Organ Pipe. In three months following a rainy period in the winter of 2004–2005, some plants in a planted orchard grew 15 cm. (Fig. 2.10.)

Recruitment of juveniles in Sonora and (to a lesser extent) Baja California is often lacking or minimal. This failure to replace older populations is often attributed to overstocking of livestock. Cows tend to trample the young plants, and goats eat them. Both remove the vegetative cover that apparently protects the seedlings from predation by ants and from sunburn. Successful germination also appears

Figure 2.9. Organ pipe covered with ballmosses (Tillandsia *spp.), Sierra San Francisco, Baja California.*

to require a time of extensive rainfall associated with seed set for the seedlings to become established.

The variables of stocking rates and the timeliness of rainfall are not the sole factors in the appearance of young pitayas, however. Although recruitment is near zero in large portions of the dense pitaya forest of southern Sonora, even within fenced areas from which livestock is excluded, young plants seems to be doing fine in two widely separated areas, both of which are subject to grazing. In one, near the coast, large numbers of juveniles of a wide range of ages dot the desert floor, many of them exposed to nearly full sunlight. In another, a heavily trampled thornscrub location near a village, the existence of large herds of burros, cows, and goats has not prevented the germination of hundreds of young pitayas. The reason for the discrepancies is yet to be discovered. Just why millions of pitayas germinated and grew to old age in a large area where no young ones are now to be found is a mystery. Botanist Mark Dimmitt

Figure 2.10. Experimental orchard of pitayas (organ pipe cacti), Sirebampo, Sonora.

has suggested to me that competition with adult plants for water accounts for the absence of young plants in a pitaya forest. (Fig. 2.11.)

The organ pipe is closely related to two other columnars, *S. martinezii* and *S. quevedonis,* and perhaps to a shrubby semicolumnar, *S. beneckei.*[19] These plants have gummy areoles called "glandular trichomes" on new growth that exude a saplike substance that dries hard and darkens to reddish or brown. They also exhibit a horizontal notch or crease between areoles on some ribs, but not on all. This feature may be lacking or spotty on some branches or even on some individuals but is nearly always present somewhere on the plants. (Fig. 2.12.)

A crested form of the pitaya (*aaqui nábera*) pops up sporadically throughout the region. It often yields no fruit. When it does, some traditional Mayos warn that the fruits should not be eaten. The plant is bad, they say,

and the fruit is thus contaminated. An apparent hybrid form of the pitaya, a cross with sina (*Stenocereus alamosensis*), is found throughout the Comunidad de Masiaca in southern Sonora. It is a columnar cactus up to 5 m tall, with curling branches, but with reddish pink instead of white flowers. The fruits are excellent, similar to those of the sina, and the dried branches are hollow, the wood hard and durable. Preliminary studies by Mexican ecobiologist Alberto Búrquez Montijo suggest that the hybrid may be fertile. (Figs. 2.13 and 2.14.)

Pitayas are pollinated by bats, hummingbirds, and a grab bag of smaller foragers.[20] Flowers are white (sometimes pinkish), and they open at night. They may remain open well into the following day, even as late as the afternoon. The flowers and subsequent fruits develop along the upper parts of the ribs, usually within 1 m of the apex and well out of the easy reach of most human collectors. (Fig. 2.15.)

Figure 2.11. Juvenile organ pipe cacti in full sun, Bajerobeta, Sonora. In the photo are eighteen pitayas less than 50 cm (20 in.) tall, plus the three held by the person.

Figure 2.12. Branch tip of Stenocereus thurberi. *Notice the notch between areoles.*

Figure 2.13. Aaqui nábera (crested organ pipe), Masiaca, Sonora.

Figure 2.14. Sinaaqui, *hybrid, probably between* Stenocereus alamosensis *and* S. thurberi, *Masiaca, Sonora.*

For most native peoples of Sonora, the pitaya is the most important native plant. It is also the most noticeable plant in the vegetation of coastal and foothills thornscrub communities. In thornscrub, it often appears to overwhelm the landscape. In other areas, it is a codominant with scrubby leguminous trees. Pitayas populate the coastal plain south and east of Huatabampo to the Sinaloa border in dense forests. Huge swathes, however, once thickly populated with pitayas, have been eradicated by commercial agriculture and shrimp farms. From Sinaloa southward, the pitaya is found only sporadically until it vanishes roughly 200 km south of the Sinaloa-Sonora state line. Rainfall slowly increases to the south of the Sinaloan border, and the increasingly richer thornscrub apparently begins to crowd out young pitayas. Even so, pitayas grow in sufficient numbers in coastal northern Sinaloa that harvesters migrate from the foothills during August to pick and market the fruits. Somewhere in the latitude of Culiacán, pitayas disappear, and *S. martinezii* appears. The latter has a noticeable trunk.

Where trails make passage possible, a walk through the dense growth of pitayas is an incomparable experience.

Figure 2.15. Pitaya (organ pipe, Stenocereus thurberi*) flower, Masiaca, Sonora.*

Figure 2.16. Baby pitayas (organ pipe cacti), Sirebampo, Sonora. They are hawked along the International Highway.

In places, more than six hundred adult plants per hectare can be found. This habitat, known locally as El Pitayal (vegetation dominated by pitayas), is a national treasure of Mexico, a spectacular landscape in which a multitude of angular arms seems to crisscross all available space. This poorly recognized forest also produces tons of sweet fruits.

Flowering of most pitayas begins in late spring, earlier in some plants and some years. By July, the earliest fruits ripen, but serious collecting seldom begins prior to August, when hosts of ripe fruits appear and collecting becomes worthwhile. In 2005, the fruits began to ripen in early June. Immature fruits (*caboasi*, "it is still green") are inedible. *Poposáhuim* ("between green and ripe") are better. Pickers harvest them and leave them to ripen in a bucket. When fruits are ripe *(huásim)*, vast numbers never make it home.

In Baja California, where the cactus is also common, the fruit harvest was so vital to indigenous Californians' survival that skirmishes broke out among groups competing for groves of pitayas. The various growth stages of the pitayas were the basis for the annual calendar of several native peoples of Baja California, the new year beginning with the first pitaya fruits in late June or early July. Fr. Miguel del Barco, writing about Baja California in the mid–eighteenth century, observed of the fruits that "those that have a yellow skin, some have white meat inside, others yellow, and still others buff. All of them are excellent fruit, worthy of the table of the greatest monarchs."[21] One Jesuit missionary fumed about the natives' fondness for the fruits:

> In the immediate vicinity of our camp were many of the tart pitaya, the only thing that, throughout the Californias, might be termed a luxury. These were coveted by the Indians; no matter what orders were issued by the captains, and, despite all I could say, they would not restrain themselves. Whenever they went out, hatchet in hand, after wood, or sought water or anything else, they invariably strayed away. So irremediable was this evil that it was a strong temptation to wish that the Californians had never acquired this habit.[22]

Hia-Ced O'odham, Pimas, Seris, Tohono O'odham, and the now extinct Ópatas also relished these fruits, which inspired such reckless and wanton disobedience among Baja California peoples. An older Mayo once mentioned to me that pitaya fruits have aphrodisiac powers, and it is this belief, apparently common, as well as the knowledge that the fruits could be used to brew a mildly alcoholic wine, that may have struck primordial terror into the hearts of the Jesuits, who were instructed strictly to forbid the drinking of wine among their native charges.[23]

Pitayas have played an important role in the lives of Cáhita people—Mayos and Yaquis. Their languages, linguistically close, contain many terms for aaqui, the cactus, and its parts. Because pitayas grow in greatest density in Mayo country, I describe some of their uses and their terminology as I learned them in a couple of decades in the field with Mayos.

In Mayo and Yaqui country, fruits *(aaqui tej'ua)* ripen in immense numbers in late July through September, forming an important component of the Mayo and Yaqui diet and a potentially valuable economic resource. Mayos point out that one never gets tired of eating them. They consume as many as fifty a day. The fruits were also formerly dried and preserved or made into wine, although this practice seems largely to have died out. Unlike the saguaro, pitaya fruits of the region do not dry within the husks while still on the plant (at higher altitudes they may), perhaps because of the higher water content in the fruits, higher humidity in the pitaya's range, and more abundant animal consumers. (Fig. 2.16.)

Fruits ripen earlier, usually by early July at the northern parts of the pitaya's range and by mid- to late July farther south. The railroad town of Carbó in central Sonora is home to the largest and reputedly tastiest pitayas. They mature about a month earlier than those farther south in Mayo and Yaqui lands.

Pitaya seca (dried organ pipe fruit) is also prepared in widely scattered locations. It is usually made in small batches of a couple of dozen ripe fruits that are boiled in a skillet with a little water. The boiled mixture *(beja buasic)* is strained through a coarse piece of cloth to remove excess *miel* (syrup). The remaining mass, rather slimy, is spread out to dry, covered with a screen to keep out flies and insects. Adequately dried, the pitaya seca will remain edible for several months. Older Mayos still use a raised bed called a *tapanco* constructed from slats of dried ribs of the pitaya cactus laid side by side and close together to store the drying fruit and protect it from dogs, hens, and pigs.

In spite of the intense heat and humidity, pitaya-gathering time is one of the happiest of the year for Mayos. Collectors of both sexes fashion a collecting spear called *bacote* in Spanish and *jíabuia* in Mayo, usually a *quiote* (flowering stalk) of an agave or, if that is not available, the ribs of pitayas or etchos lashed together. On good collecting days, the poposáhuim fruits are collected in large plastic buckets. They are left till late afternoon or overnight, during which time the spines soften, making peeling less hazardous to the fingers. (Fig. 2.17.)

Fruits gatherers usually consume a large number of the fruits on the spot. Families who are harvesting the fruits to

Figure 2.17. Organ pipe cactus buds, Masiaca, Sonora.

sell in regional markets eat only damaged fruits, reserving the best for sale. Most pickers dexterously despine the fruit without being pricked by the thorns, which vary from plant to plant in the painfulness of their affliction. By the time children have reached ten years of age, they are usually dexterous in harvesting and eat the first few dozen fruits they collect. Around the villages, the personalities of individual cacti are well known, and those with spines that inflict the most painful punctures are treated delicately.

The quality of the fruit varies from cactus to cactus, as does the color. The dominant color of the pulp is a dark red. The best fruits are often said to be those uncommon individuals with *guinda* (purple-red) pulp. Approximately 5 percent of the plants yield a whitish pulp called *zarca* (Spanish) or *tótosi* (Mayo) with a more delicate flavor. A pitaya occasionally (I have seen only three such plants)

Figure 2.18. Organ pipe fruits, Sirebampo, Sonora. Natives consume them in great numbers.

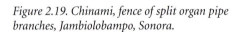

Figure 2.19. Chinami, fence of split organ pipe branches, Jambiolobampo, Sonora.

yields fruits with yellow pulp. I tried one and found it to have a more delicate flavor than the normal dark red variety. An extremely rare color is *lila,* an attractive bright purple. Each plant apparently bears the same color of fruits from year to year.

The number and size of fruits borne on the cacti varies considerably from year to year. Native observers point out that good summer rains will result in larger fruits and a longer fruiting season, while drought years result in fewer and smaller fruits. In the summer of 2001 following a nearly rainless spring and a dry previous summer, large portions of the pitaya forests bore few, if any, fruits.

Natives of the region gather the fruits and sell them in local markets. Street vendors often hawk the fruits at traffic stoplights. In the summer of 2005, the fruits sold for two pesos (twenty cents) each. In each village, several people are employed during the harvest season to gather the fruits, often selling them to a local buyer, who in turn markets them in the cities. For several hundred individuals in Sonora, collecting and selling pitayas is a vital source of income. In Carbó alone, fifty families rely on the income from the sale of pitayas to sustain them through the hot summer months. (Fig. 2.18.)

In addition to its fruits, the pitaya is a most versatile plant. The ribs of the pitaya branches, dried or green, are commonly used for fences *(chinámim).* The dried and split arms, when woven among strands of wire, provide a dense fencing that is attractive and affords privacy and protection.

Branches and ribs, whole or sectioned, are incorporated into walls, ceilings, and even some furniture. Along the coastal plain from Huatabampo south into Sinaloa, the walls of many homes are made of woven pitaya arms, harvested green. The wood of the trunk and arms is surprisingly sturdy and is used in many phases of building, including crossbeams bearing heavy weights. Lower wood is used to form the wooden base in saddles and packsaddles *(fustes).* Prior to the mid-1950s, when barbed wire became available and affordable in rural Sonora, large numbers of cuttings were planted in rows to produce living fences, many of which can still be seen in the region. The dried wood is used in great quantities as fuel in ovens for baking bread and in kilns for firing burnt adobes. Andrés Pérez de Ríbas, writing in the mid-seventeenth century, observed that on chilly winter nights, natives of the region would gather ribs for heating.[24] (Figs. 2.19, 2.20, and 2.21.)

The plant has medical uses as well. The fleshy, moist stem is singed to remove spines and then applied to the flesh directly for snake and insect bites. Other Mayos reported that the scorched peel of the fruit is applied directly to the anus for hemorrhoids, cautioning that it must be scorched enough so that the spines are burned off. In one village, dried peels *(aaqui begua)* are boiled into a tea and taken to get rid of stomachache and to alleviate hemorrhaging in women.

An odd use of the cactus came to me through a Mayo friend in southern Sonora. He noted that the smoke from

Figure 2.20. Ceiling beams of organ pipe branches, Loma de Angeles, Sonora.

Figure 2.21. Fence and home of split pitaya branches, Huebampo, Sonora.

burning pitaya branches would neutralize the musk sprayed from a skunk. His procedure was to build a small fire of the ribs in one room, let it smolder, and shut in the dog that had been sprayed. He swore that the smoke, if it reached all the affected areas of the animal, would eliminate the powerful stench of skunk musk. Pity the dog.

Felger and Moser report that Seris reportedly used dried fruit castings mixed with animal oils to produce a caulk for patching and filling spaces and holes in seagoing boats.[25]

A solitary pitaya is probably the most venerated plant in all of northwestern Mexico, perhaps the entire republic. The 3-m-tall plant, said to be more than 250 years old, juts out from ancient rockwork high in the side wall of the church in the former mining town of La Aduana, Sonora, near Alamos. As the story goes, the cactus sprouted from the south wall of the church shortly after it was completed in 1730, and the Virgin Mary then appeared among the branches. Pilgrims from all over Mexico visit the plant during the November 20–21 festival of the Virgin of Valderrana. Many of the faithful leave votive candles behind. In the mid-1990s, the rivulet of wax from thousands of candles caught on fire and singed the bottom half of the famed pitaya. Despite great concern for the plant's survival, new growth soon appeared, and the bright green color proved that the plant was far from dead. (Fig. 2.22.)

In spite of the enormous numbers of pitayas in Sonora, threats to the densest populations include clearing for commercial agriculture and shrimp farming and clearing plus the introduction of fire-loving range grasses such as buffelgrass (*Pennisetum ciliare* [L.] Link). Already nearly half of the densest pitayal has been cleared, most of it since the 1980s. Unfortunately for the pitayas and the native peoples who use them, the agricultural and aquaculture schemes proposed in the region place little value on the pitayas and their habitat. (Fig. 2.23.)

The Aztecs, their contemporary southern neighbors (especially the Mixtecos and Popolocans), and the long succession of their ancestors recognized the value of pitayos and pitayas, celebrating the virtues of the plants and their fruits in myth, song, and codex. Today the delectable fruits are still harvested, but in even greater abundance and succulence: descendants of the pre-Columbian groups and their Mestizo progeny have selected the best fruit producers from wild cacti, planted cuttings, then planted cuttings of the best plants grown from these cuttings. By this gradual selection, they have come to produce large fruits (up to 9 cm in diameter) renowned for their delicate flavor and juicy texture. The genus *Stenocereus* seems to lay claim to most of the best fruit producers among cacti, for it is the only genus of columnar cacti whose members are widely

Figure 2.22. Organ pipe growing from church wall, La Aduana, Sonora. Pilgrims arrive from throughout Mexico to pay homage to this plant. Supplicants have left the ribbons.

cultivated and whose fruits are produced on a commercial scale. At least six species have become horticultural crops throughout central, northwestern, and southwestern Mexico. In various locations, farmers have planted orchards, sometimes of one species only, often of mixed varieties. The plants and their fruits have been an important constituent in local diets for millennia. As early as 5400 years ago, the fruits, seeds, and stems were part of the regional diet, and 3500 years ago the pitayas may have constituted the most important plants in human diets.[26]

Figure 2.23. Inside Organ Pipe Cactus Reserve near Coteco, Sonora.

The Sahuira

Stenocereus montanus (Britton & Rose) Buxb.
sahuira, pitaya colorada, pitaya sahuira

Sahuiras are little known outside their habitat on Mexico's West Coast—southwestern Chihuahua, southern Sonora, Sinaloa, and reportedly Nayarit, Jalisco, Colima, and Zacatecas. Due to their ecological requirements of mountainous terrain and their scarcity near urban centers, they are usually given short shrift in discussions of the genus *Stenocereus,* yet in their habitat they are a key plant species for those who live near them, providing fruit, lumber, and medicine to the Mayos of the Río Fuerte and to many other people as well. In the mountainous country above the coastal plains of extreme southern Sonora and the foothills in northern Sinaloa, *S. thurberi,* so common on the coastal plains 50 km to the west and north, becomes rather uncommon, and sahuiras dominate.

These estimable, big plants commonly reach 10 m in height (13 m rarely) in well-developed tropical deciduous forests—from 200 m elevation near El Fuerte to nearly 600 m in Sonora. The plants (and the fruits) tend to be rather larger than their pitaya (*Stenocereus thurberi*) cousins but never occur in the latter's enormous numbers. Plant explorer Howard S. Gentry estimated one plant growing in close forest at 20 m tall. Sahuiras superficially resemble the sympatric etcho (*Pachycereus pecten-aboriginum*) in that they have a well-developed trunk and arms that issue from the trunk well above the ground. Natives who harvest the fruits report that the plants are prolific fruit producers, averaging in excess of three hundred fruits per plant, often many more. Some report that although most plants fruit in June, individual plants will produce fruits well into September, giving the species one of the longest fruiting periods in the genus. Spines on the fruits often take on a reddish hue, making them look rufous when backlit, hence the common name *pitaya colorada.* (Figs. 2.24 and 2.25.)

Figure 2.24. Sahuira (Stenocereus montanus) *near El Chinal, Sonora.*

Figure 2.25. Sahuira near La Aduana, Sonora.

The northernmost sahuiras grow in extreme southern Sonora in the Río Fuerte drainage on hillsides above Alamos, and in the foothills of the east bank of the Río Mayo. Those in the Río Fuerte drainage of Sonora usually grow in association with *Pachycereus pecten-aboriginum* and other trees of the well-developed tropical deciduous forest, such as *Bursera grandifolia, B. stenophylla, Ceiba acuminata, Chlorophylla tinctoria, Cochlospermum vitifolium, Ipomoea arborescens, Lysiloma divaricatum, Pseudobombax palmeri,* and a host of others. In granitic soil near the ancient Mayo village of Baca on the southeast bank of the Río Fuerte in Sinaloa, an enormous individual on an outcrop has grown in the form of a candelabrum with more than fifty upright arms. It casts a considerable and comforting shadow. (Fig. 2.26.)

In mountainous northern Sinaloa, the sahuira supplants the pitaya in many domestic uses, providing lumber and ample fruits. The technique for gathering fruits and consumption are similar to those for *S. thurberi*. Harvesting the sahuira fruits requires a certain amount of experience and delicacy, for the fruits are covered with spines even longer than those of pitayas. Clumsy gathering can be a painful experience. If the fruits are ripe, however, within a few hours of harvesting the spines usually drop off or are easily scraped off.

Sahuiras crop up early in the recorded history of northwestern Mexico. The intrepid Spanish warrior Captain Diego Martínez de Hurdaide and his troops conquered the region in 1605. At that time, the region was occupied by several peoples (including Basiroas, Huites, and Zuaques) who spoke variants of the Cáhita linguistic group, the same family of languages spoken by Mayos and Yaquis. The early Jesuit Pérez de Ribas reported that they, as all lowland peoples of the Northwest, relished the fruits of the pitaya. He understandably did not distinguish among the various species of pitayas. In the case of the Huites and Zuaques, the "pitaya" fruits would have been sahuira fruits. As the various groups came under Spanish domination, the Spaniards and Jesuits brought them into larger mission towns, where their spiritual and moral development could be carefully scrutinized. During the many rebellions and hostilities, peoples of different ethnic derivation came together, voluntarily or involuntarily, as refugees, indentured servants, prisoners, or even converts, to live in the mission towns.

Figure 2.26. Sahuira near Baca, Sinaloa.

Over the centuries, all the groups in the Río Fuerte region came to be known as Mayos, thus glossing over their varied cultural backgrounds. One important thing they had in common (in addition to mutually intelligible languages) was their widespread use of the sahuira.

People and all manner of wild beasts eagerly eat the fruits, locally called pitayas. Mayos of Los Capomos, Sinaloa, pronounce them to be unquestionably superior to the fruits of *S. thurberi*. I like both, but sahuira fruits are somewhat larger and easier to harvest. The most common pulp color is red, but orange and white fruits are not uncommon. I have not tried these other flavors. For the region upstream of El Fuerte well into Chihuahua, sahuira fruits are the most important cactus fruit. In all probability, the regionally famous June 12 Fiesta de San Pedro in Los Capomos, a Mayo community northeast of El Fuerte, represents a modern continuation of the aboriginal celebration of the ripening of the sahuira fruits. During this regional celebration, thousands of pilgrims assemble in the town to watch Mayo dancers—*pascolas* and deer dancers—perform to the music of renowned musicians from all over the region and to eat sahuira fruits.

Mayos from Los Capomos report that wine was formerly fermented from the fruits. Nowadays no one makes it, they lament, because commercial booze has replaced it. But some Mayos gave me assurances that it could be done as a demonstration. At times, local people will dry the fruit for preservation, but the climate in the mountains, considerably more moist than in the flats below, makes drying more difficult.

Natives of the region say that sahuira wood is more durable than that of the sturdy etcho. It can reportedly be worked as fine hardwood for making furniture. Both the arms and the base are sources of lumber. At Los Capomos, natives use the wood to make beds, doors, and tables, as well as vigas and *horcones*. At El Chinal and Güirocoba, Sonora, north of the Río Fuerte, cuttings of sahuiras in combination with those of *S. thurberi* and *Pachycereus pecten-aboriginum* were planted many decades ago as part of living fences. In Sinaloa, fences are often composed of sahuira exclusively. Cuttings of the plants are placed side by side in rows and grow into palisadelike fences that can withstand the attacks of motor vehicles. The natives of indigenous communities have cultivated the plant for fruit, medicine, and lumber for so long that most homes have a fence or window of sahuiras raised from cuttings, and many country lanes are bordered by well-tended rows of this worthy plant. Since the introduction of barbed wire in the 1950s, living fences have not been widely planted. Orchards of

Figure 2.27. Table made of sahuira wood, El Chinal, Sonora.

the plants can still be found in foothill communities in the Río Fuerte region. (Fig. 2.27.)

A gentleman from Baimena, a Mayo indigenous community, reports that each year he saves a supply of husks from the fruits, singes them, and keeps them as a remedy for hemorrhoids, applied directly to the anus. Others report that the fruits themselves are eaten to cleanse the insides and to cure ulcers. This remedy is far more pleasant than most medicines and vastly more desirable than injections. We should hope that all medicines were so agreeable.

Guarijíos of the lower Sierra Madre of Chihuahua and Sonora reported to me that they make tamales from the mashed seeds of the sahuira, which they call *sahuí*. The plant is nowhere common in their territory, and they must walk considerable distance over rugged terrain in the hot weather of June to gather the fruits. The uniform excellence of the sweet fruits makes the effort well worth their time, they assured me. Unfortunately, they explained all this to me during the winter months when no fruits were available, so I have not been able to watch the process or sample the product. The etcho is the most important plant for the Guarijíos, so the sahuira fruits must be delectable indeed.

Sahuiras and etchos are easy to confuse. Sahuiras are of a lighter green color than etchos. Their characteristic appearance more resembles a candelabrum than the tight formation of the etcho arms. Sahuira areoles are raised and, though often reddish when new, generally turn dark or black, the result of a hardened gummy exudate (areolar or glandular trichomes), whereas etcho areoles are nearly flat and somewhat felty, gray in color. Sahuiras share the black trichomes with *Stenocereus chrysocarpus* and *S. queretaroensis,* both large columnars of Nueva Galicia.[27] The sahuira's needlelike spines are brown to black, whereas the etcho's spines are gray to white and are thickened at the base, more reminiscent of thorns. The arrangement of the etcho's sturdy spines give the ribs a decidedly more bumpy appearance than that of the sahuira, whose delicate spines are less visible, giving the plant the appearance from a distance of having smooth ribs. Sahuiras have fewer ribs, seven to eight as opposed to nine to eleven for etchos. The furrows between the ribs on etchos are deeper and more pronounced. (Fig. 2.28.)

Sahuira flowers open at night[28] and closely resemble those of the pitaya *(Stenocereus thurberi);* the fruits are characteristic of *Stenocereus,* with visible areoles from which several needlelike spines emerge. Sahuira fruits have fewer spines than pitaya fruits, but the spines are sharper than those of the etcho. The husk is clearly visible beneath the spines, whereas on the etcho it is buried beneath the thick coat of rather weak spines. The pulp of etcho fruits is characteristically reddish purple, whereas that of sahuira fruits varies from yellow to red.

The Pitayos of Nueva Galicia

Nueva Galicia is the name given by Spaniards centuries ago to southwestern Mexico, the area now included in the states of Colima, Jalisco, Michoacán, and western Guerrero, apparently because the plateau portions of the region bear striking similarity to Spanish Galicia. There is no sharp political or geographical point setting the region off from the western part of Mexico, but for our purposes the regional denomination works well. One species of *Stenocereus* is found growing wild in the temperate central uplands of Jalisco *(S. queretaroensis),* and four or five species grow in the torrid lowlands.

Stenocereus queretaroensis (Weber) Buxbaum
pitayo, pitayo de Querétaro

The *pitayo de Querétaro*—or simply *pitayo* or occasionally *pitayo blanco* (the fruits are called *pitayas*)—is common in west-central Mexico, favoring mesic hillsides and bajadas from Querétaro west to Jalisco, north through Zacatecas, and south to Colima and Michoacán. I believe I have spot-

Figure 2.28. Close-up of juvenile sahuira. Note reddish spines and gummy areoles.

ted it on hillsides in Hidalgo and in the state of Mexico as well. It probably once grew in the deep-soiled valleys as well, but these valleys were converted to agriculture many centuries ago, and virtually none of the original vegetation remains. The plant seems far more common in Jalisco than in Querétaro, but its species name still sticks. Large specimens exceed 10 m in height, more than 33 ft., with dozens of branches, but most plants are smaller. They often assume candelabra form, with many arms rising from a single trunk. As is often the case in this genus, it is a prolific fruit producer, and the branches are often heavily laden with developing fruits in April and early May. The fruits grow along the sides as well as at the apex of the branches. When the fruits are ripening, their bristly and spiny husks make the plants appear from a distance as though they are laden with Cyclops or one-eyed Muppets, rather like the fruits of *Pachycereus pecten-aboriginum* and *P. grandis.* In April and May, the ripening fruits turn reddish and purplish and are gathered and marketed, though many of the

Figure 2.29. Pitayo de Querétaro (Stenocereus queretaroensis). Although many thousands of these plants are cultivated, this one is wild.

fruits never make it to the market because they are eagerly consumed by the harvesters as well as by birds, insects, and mammals. (Fig. 2.29.)

S. queretaroensis is neither the biggest nor the most attractive of the columnar genus *Stenocereus.* It is not especially tall, nor does it grow naturally in dense forests (although it may have before its habitat was transformed into a human-shaped landscape). Its fruits, however, have taken to domestication (or semidomestication) like few others. In the Valle de Sayula in Jalisco, cultivating plantations of pitayos for their fruit has become a successful commercial enterprise—more than 1000 ha of the plants are in cultivation.[29] The average weight per fruit of the favored mamey variety is 165 g (more than ⅓ lb.), three times the average for fruits of *S. thurberi,* and a good plant will yield five hundred fruits. The fruits have the added advantage (along with several other *Stenocereus* species) that the spines slide easily from the husks when fruits mature, making harvesting far easier and less hazardous than for most other cactus fruits. In the municipality (county) of

Techaluta, Jalisco, orchards—often plots of a tenth of a hectare—of carefully tended cacti grow within the village and beyond and have spread into nearby villages as well. Many backyard plants are harvested commercially. In the village of El Poste, densely planted orchards produce yellow and white fruits; the orchards of Techaluta are better known for reds and purples. (Fig. 2.30.)

As a result of selection, at least seven different varieties of fruits, producing a rainbow of colors and varying degrees of sweetness, have been selected over the centuries. The largest fruits, called mameyes, are worthy of praise, indeed, and bring top prices in nearby Guadalajara.

Techaluta townsfolk claim with some justification that pitayas (fruits of the pitayo) from Techaluta are the best in the world, a claim that would be disputed by Oaxacans and by Sonorans from Carbó. Techalutans each May celebrate a fiesta honoring the pitayo and its harvest. The fiesta has become the social highlight of the year in the municipality. A pitaya fruit cooperative in the town now boasts a mechanized system for producing pitaya jam, a gift from the federal government that saw great potential in the local enterprise (but that lies unused for ten months of the year). The tennis-ball-size fruits are marketed regionally and bring premium prices—five pesos each (fifty cents U.S.) in the year 2000 at the height of the season. (Figs. 2.31a and 2.31b.)

The commercial success of the plantations has been remarkable, so much so that Mexican government investigators are researching ways to promote horticultural development of pitayo plantations elsewhere. The Techaluta producers claim that the economic return from the pitayos is greater than for any competitive crop, even though the harvest season is hardly eight weeks in length. The orchards are usually seeded with a forage grass and double as pastures. Cattle appear to present no harm to the pitayos once the plants have achieved sufficient size (cows tend to trample and gnaw younger plants). In some cases, the cacti rise from fields sown with corn and beans. In other cases, they form fences or boundary markers. (I wonder if landowners collect only the fruits on their side of the "wall"?) The cacti are trimmed to keep them of manageable height because fruits on plants more than 3 m tall may become inaccessible to harvesters. Wild cacti (or planted cacti allowed to reach normal height) are often found adjacent to cultivated rows and may be important for cross-pollination.

Local cultivators report few management problems. Colonies of ants from time to time attack the growing tips of plants, inflicting considerable damage and lowering crop yields. Some farmers respond by applying insecticides

Figure 2.30. Orchard of Stenocereus queretaroensis, *Sayula, Jalisco. The plants produce hundreds of fruits and require little maintenance.*

Figure 2.31a. Pitayeros *(pitaya harvesters), Amacueca, Jalisco.*

Figure 2.31b. Ripening pitayas, Techaluta, Jalisco.

around the base of the plants. The larvae of a beetle called *plaga de pitayas* apparently can cause damage to significant numbers of fruits, but appear only sporadically. Applications of insecticides can be effective but are expensive and can be difficult and hazardous to apply, so their use is not common. Most of the labor for harvesting is local, so considerable care is exercised in applying toxic chemicals to the plants. Poisoning one's neighbors is a bad policy.

Production around Techaluta is expanding as farmers plant cuttings in increasing numbers of their plots. Propagation is simple. Arms of the parent plants are lopped off, preferably during the dry season just before the arrival of the summer rains. They are allowed to dry for a couple of weeks. The cut end is simply placed in the ground like a fence post. One farmer reported that he plants arms immediately after a rain to assure moisture for the newly sprouted roots. These cuttings sprout quickly, but still must grow for four to five years before they begin to produce on a commercial scale. One producer claimed that in only two years his cuttings begin to produce, the plants growing more than 20 cm each year. Cultivators for the most part do not believe supplemental irrigation increases yield of pitayas. Some suggest that fertilization actually decreases pitaya production, even though some studies suggest that adding manure around the base of the plants encourages plant growth. Several producers mentioned that fertilizers interfere with fruit production. One lamented that he had attempted to boost production by adding goat dung around the base, but the plants grew when the fruit should have been growing, and the fruits fell off the plants. The rapid growth of the plants requires that some branches be pruned each year to provide sufficient space for developing fruits and for harvesters to pass between and among the plants.

These pitayos appear to need a frost-free environment (occasional frosts damage plants and may damage buds), rainfall in excess of 750 mm (the rainfall at Techaluta is around 800 mm, according to local producers), and the moist summer rainy season beginning in late May—requirements that will considerably restrict commercial applications in the more arid parts of Mexico. For example, rainfall is less bountiful in the Sonoran Desert, where *Stenocereus thurberi* grows, usually below 400 mm. The wild pitayas there are smaller, their size more variable, and production is less reliable.

In Techaluta, other producers have experimented in making a *vino,* or alcoholic beverage, by fermenting the fruits. A sweet from the seed dried and pressed into cakes is sold commercially. Pulp from fruits that lack commercial quality is heated and poured into trays and dried. This is sold as pitaya seca and can be reconstituted by soaking. One family is experimenting in the production of commercial shampoo from the pitayo branches.

Stenocereus chrysocarpus Sánchez-Mejorada
pitayo, pitayo pachón

S. friçii Sánchez-Mejorada
pitayo de aguas, tambar

S. quevedonis J. G. Ortega
pitire

S. standleyi (J. G. Ortega) Buxb.
pitaya, pitaya marismeña

Three species of *Stenocereus* are primarily (though not exclusively) found inland, beginning around 50 km from the coast of Nueva Galicia. *S. quevedonis* and *S. friçii* are especially common in the lengthy and hot, deep valley and canyons of the Río Balsas in Guerrero and Michoacán and in the basin of a principal Michoacán tributary, the Río Tepalcatepec. This region is perhaps the hottest in central Mexico. The vegetation ranges from scrubby to well-developed thornscrub, where cacti are the tallest plants, to tropical deciduous forest, where the canopy of drought-deciduous trees exceeds the tops of the tallest columnar cacti. (The lower Balsas basin should be considered the *Stenocereus* capital of the world.) *S. standleyi* grows on cliffs and steep slopes along the coast throughout the region, but it has become rather uncommon due to habitat loss.

Much of this rugged and hot portion of Mexico is (or was until recently) sparsely inhabited except for occasional towns adjacent to well-watered drainages. Its flora and vegetation are also rather poorly studied, and several conspicuous columnar cacti in the region were not described until the latter part of the twentieth century. Both the Río Balsas and the Río Tepalcatepec drain well-watered basins that appear to form a distinct region. The Río Balsas basin, Mexico's largest, is a tectonically active depression and includes a major portion of the state of Guerrero and parts of Michoacán, Mexico, Morelos, and Oaxaca as well. The Río Tepalcatepec empties into the Balsas from the north and west, draining much of southwestern Michoacán and southeastern Jalisco. The intensely warm climate and the topographical isolation of these basins explain the appearance of several endemic columnars, including *Backebergia militaris, Pachycereus tepamo, Stenocereus friçii,* and *S. chrysocarpus.* The lower Balsas basin is host to at least eleven species of columnar cacti, fewer than the array found in the Valle de Tehuacán of Puebla, but certainly a remarkable example of cactus diversity.[30] If we include upstream tributaries (such as Cañon de Zopilote

*Figure 2.32. Pachón
(Stenocereus chrysocarpus)
near Palo Pintado, Michoacán.
Both the fruits and the lumber
of this narrowly distributed
species are used locally.*

in Guerrero), the number increases to at least fourteen. Most biogeographers consider the Balsas and Tepalcatepec basins to be distinct biogeographic provinces of Mexico. If endemism of columnar cacti is an indicator, they are surely correct. Even better for the cactologist, large portions of the basins retain their native vegetative cover. Much of the middle Tepalcatepec basin has been the subject of intense agribusiness development, however.

Of the three large endemic or quasi-endemic *Stenocereus* columnars, *S. chrysocarpus,* the *pachón,* appears to have the most restricted range. It has been reported previously only from in and near the Río Tepalcatepec basin, where it grows on calcareous soils. It is quite common around the villages of El Espinal and Palo Pintado on Highway 37. Some 10 km south of Arteaga on cleared hillsides, it stands out handsomely. Although I have not found previous reports from the slopes adjacent to Presa Infiernillo, I was surprised to discover it doing quite well on some steep

northern slopes there, where it flourished in inhospitable tropical deciduous forest. I also found six large and healthy plants growing on a densely forested knoll 15 km east of the Nahua village of Maruata on the Michoacán coast, far from the limits of the Infiernillo. It is easy to see why this cactus had not been previously reported, for it was not obvious among the grays and browns of the winter vegetation and would be well nigh invisible in the green of summer. I suspect that further investigation will reveal that this most noticeable plant has a much wider range than previously thought, so its supposed restricted distribution may simply reflect inadequate exploration of the region. This lack of field collections is understandable, for the climate is hot to beastly hot, and the hills are rugged and thickly vegetated with very thorny bushes and trees that make plant explorations challenging indeed. To take to the hills in the region requires the services of an adroit *machetero* who knows his way around. (Fig. 2.32.)

The pachón, as *S. chrysocarpus* is commonly called, was described by the Mexican cactologist Hernán Sánchez-Mejorada R.[31] He was struck by the handsomeness of the fruit husk, whose yellow spines seem to shine—hence the species name *chrysocarpus,* meaning "shiny fruit." This expansive and large columnar cactus reaches in excess of 10 m in height and is usually a *shiny* light green in color and exhibits only slight ribs, making it distinctive in a region of many columnars. The branches are quite succulent; the ridges are not particularly high; and the spines tend to be short, so the waxy cuticle reflects light well. The pachón has six to eight ribs, usually seven. The areoles are dark to black and grow quite close together, usually no more than one-half inch apart, another distinctive feature. The flowers, which open in February through April, are white. The scarlet pulp of the fruits, which ripen in May, is highly edible, and the fruits (called *pachonas*) can be found in markets in Arteaga and sometimes along Mexico Highway 37 in the region. The ripening fruits are covered with a menacing shield of hundreds of sharp, bright yellow spines, but this shield sloughs off as a unit when the pulp is ripe, making a most interesting and attractive, if potentially damaging decorative knickknack as it lies on the ground. Natives of the region remark that the fruits continue to ripen until the rains arrive in late May. With the additional moisture, any fruits remaining on the plants rot, and so ends the fruit production for the year. When I first visited the area in July 2001, the rains had not begun, and dried fruits remained on a few of the plants.

The wood from the pachón branches is also widely used in the region for ceiling beams, or vigas. Residents attribute the plant's use as lumber to its larger, wider branches, which, they say, provide stronger wood than other species in the region. (Fig. 2.33.)

In a brief reconnoiter in the bush around El Espinal, I was unable to find any juvenile plants. Local peasants reported that there are a few, but not many. This situation may be a result of the overpowering effects of livestock grazing limited only by starvation of the stock. This lack of juvenile plants, quite common in Mexico, does not portend well for the future of this most worthy cactus. Elsewhere, however, I found young plants flourishing on steep slopes where access by goats was more limited. Natives also reported that many plants can be found in the upper parts of the Río Tepalcatepec basin.

The pachón is similar in appearance and habit to *S. chacalapensis* from southern Oaxaca, though the latter is larger. When plant mappers have sufficiently explored the Tepalcatepec basin and environs (exploration is being made easier by the dismaying rate of destruction of the

Figure 2.33. Roof timbers formed of pachón wood, Palo Pintado, Michoacán.

native thornscrub and tropical deciduous forest), it will be possible to formulate hypotheses to explain the pachón's highly restricted distribution. The biology of the species is poorly known, but mature pachones are sufficiently abundant around several small villages that study should not be difficult for one willing to entertain rural conditions.

Stenocereus friçii, known as *pitayo de aguas,* was not described until 1973, an indicator of its isolation in the rugged Sierra Madre del Sur and of the often stiflingly hot climate in which it grows. It is not rare, however, for it grows in groves known as *cardonales* in the vicinity of the lower Río Tepalcatepec basin and near the Presa Infiernillo, in which it mingles with even greater numbers of *S. quevedonis,* called *pitire.* It can also be found on rolling hills at less than 300 m elevation only 30 km from the Pacific Ocean. I have seen lone plants growing on rocky knolls above the Pacific Ocean at Zihuatanejo, Guerrero. In thornscrub and in open tropical deciduous forest, it grows 4–7 m tall in the typical spreading habit of the genus. In more moist semideciduous forest, it reaches 9 m in height, but its numbers are substantially fewer. (Fig. 2.34.)

The pitayo de aguas has four to six ribs and diurnal flowers, with a shiny, darkish green epidermis. The ribs are thin and deeply recessed—up to 5 cm from base to the edge. The margins of the ribs are unequal, with bumps and constrictions growing all along the extensive edge. The constrictions may be related to plant growth, perhaps one for each year. Younger branches have numerous crinkles that grow into the valleys between the ribs, making for a rough rib surface. The spines, though numerous, are short, the maximum reaching little more than 2 cm in length. The areoles tend to be dark and large, about 2 cm

Figure 2.34. Pitayo de aguas (Stenocereus friçii), *southern Michoacán. Note the four ribs, the smallest number in the genus* Stenocereus.

Figure 2.35. Flowers and early fruits of Stenocereus friçii, *southern Michoacán.*

distant, with shortish spines, so one can harvest fruits without being threatened with long, vicious spines. The ribs tend to bend—outward or inward—and the plants thus lack the symmetry of the pachones, the pitires, and the larger pitayos. This characteristic, combined with the lesser number of ribs and accented areoles, makes the plant easily identifiable. (Fig. 2.35.)

Unlike the fruits of the pachón and the pitire, which share its habitat and flower and fruit prior to the rainy season, the fruits of *S. friçii* ripen *during* the rainy season. Early fruits may ripen by mid-June, but most mature in early July through mid-August, hence the plant's common name, which translates "pitayo of the rains." They are roundish and exceed 5 cm in diameter. Their bright red pulp is most appealing, and the fruits are sold in great numbers in local markets in the region (especially in Arteaga) and along the highways. The plant is reportedly cultivated for its fruits, one of five members of the genus *Stenocereus* to be so domesticated. Plantations are reportedly located near Guadalajara, quite outside its natural habitat.[32] I have not seen them. Natives of its habitat praise the wild plant

because sale of its abundant fruits is an important source of income for them. A resident along the Cañon Infiernillo told me that many people sell the fruits and earn important money by doing so.

The pitire (*Stenocereus quevedonis*) reportedly ranges from southern Sinaloa at the extreme north of Nueva Galicia through coastal Colima, Jalisco, and Michoacán. I have not seem them growing outside the Balsas basin. Uncommon in the northern portion of their range, they are the dominant cacti in much of the lower Balsas and Tepalcatepec basins. Indeed, a village in the Cañon Infiernillo is with good reason named "El Pitireal," meaning "Grove of Pitires." (Fig. 2.36; see also figs. 1.32 and 3.10.)

S. quevedonis is a large cactus, far taller and broader than the pitayo de aguas, often exceeding 9 m in height and reaching at least 12 m, exceeded in the genus only by *S. chacalapensis* and an undescribed species from the Oaxaca coast. Although most pitire plants are more or less symmetrical and tend toward a candelabra shape, the branches usually veer away from the vertical much more than is the case with the pachón, with which it may be

Figure 2.36. Pitire (Stenocereus quevedonis), *Palo Pintado, Michoacán. Pitires' distribution is limited, but the plants are extremely common within their range. Their fruits are marketed locally.*

Figure 2.37a. Chair made of pitire wood, Palo Pintado, Michoacán.

Figure 2.37b. Pitire pitayera and her assistant, Cañon Infiernillo, Michoacán. Pitire fruits have a slightly perfumed aroma and flavor.

superficially confused. The latter's branches tend to be parallel and closer to the tree's central axis. Many pitires have noticeable, sometimes huge trunks, reaching 2 m in height. The branches have seven to nine ribs. The areoles are usually about 2 cm apart, and the spines may reach 5 cm in length. Flowers of *S. quevedonis* appear in February and March, are nocturnal, and are white to pinkish white in color, suggesting bat pollination. The fruits, smaller than those of the pitayo de aguas, appear in a variety of colors. Red fruit is the most common, but white and yellow fruits are also well known. Less common still are purple, orange, and rose-colored fruits. All are sweet and juicy and are marketed locally in southern Michoacán. The wood of the trunk is strong and light, but that of the branches is thinner than that of the pachón and so not desirable for household construction. However, it appears quite well suited for furniture, a function it serves in the village of Palo Pintado, Michoacán. (Figs. 2.37a and 2.37b.)

From a distance, pitires bear a striking resemblance to the pitayas *(S. thurberi)* of northwest Mexico, which have roughly twice the number of ribs. In both species, areoles

near the tips of branches exude a brown saplike substance providing a distinctive golden brown color. Also, in both species a horizontal notch is found between the areoles, a characteristic they share with *S. martinezii* of Sinaloa and the more horizontally inclined *S. beneckei* of Michoacán and Guerrero. Gibson infers from these shared characteristics a close genetic affinity among the species.[33] Cornejo considers *S. thurberi* to be derived from *S. quevedonis,* the rib doubling an evolutionary adaptation of cacti in cooler climates as protection from freezing.[34] The Sonoran pitaya also produces fruits in the same rainbow of colors as the pitire.

Many thousands of the graceful pitire trees cover the hills and steep canyon sides of southern Michoacán. Indeed, pitires are among the most abundant of columnar cacti, the vast numbers in the Infiernillo region presenting a most agreeable landscape. In much of the plant's range, however, extensive grazing pressure appears to be eliminating or preventing the growth of juvenile plants, so when many of the older plants die, they will not be replaced. Recruitment is probably better on the steep, thorny canyon sides

Figure 2.38. Pitaya marismeña (Stenocereus standleyi), *Playa Azul, Michoacán. This cactus grows only on bluffs overlooking the ocean and seldom grows tall enough to be considered a columnar cactus.*

that are less accessible to livestock, but the density is not as great there, and it will take someone with the stamina and fortitude of a bull to fight the thornscrub in a search for young pitires.

The *S. standleyi* plants I have seen are not legitimate columnars. They appear to be confined mostly to the edge of cliffs and steep slopes in the immediate vicinity of the ocean, which accounts for the common name *pitaya marismeña,* or "maritime pitaya." Their habit is clearly horizontal, although there are reports of upright individuals. I have not seen the fruits, but they are said to be edible and similar to those of pitires. The plants' distribution is intermittent from Sinaloa south to Guerrero. Their proclivity for steep ocean-side cliffs often renders them inaccessible. (Fig. 2.38.)

Pitayos and Pitayas of the Mixteca and Environs

Stenocereus griseus (Haw.) Buxb.
pitayo, pitayo de mayo

S. pruinosus (Otto) Buxbaum
cuepetla (Mixtec), pitayo de mayo, pitayo de octubre, *ndichicuá* (Mixtec)

S. stellatus (Pfeiff.) Riccob.
xoconochtli, *ndichicuá* (Mixtec)

S. treleasei (Vaup.) Backeb.
pitayo, tunillo

S. zopilotensis Arrela-Nava and Terrazas
pitayo

The first four of these five species are excellent producers of fruits, each with a different geographical distribution and time of production. *S. treleasei* grows only in the central valley of Oaxaca. Its native habitat is unknown, and it is cultivated only sporadically and accidentally there and elsewhere. Both *S. pruinosus* and *S. stellatus* are natives of the temperate valleys of Puebla and Oaxaca and, perhaps, of Guerrero and Morelos. Disjunct (apparently entirely cultivated) populations of *S. pruinosus* are reported from Hidalgo and Querétaro. Both species are extensively cultivated. *S. griseus* has a wide distribution in southern Mexico and the Caribbean and is cultivated in its own right, especially in Veracruz. It grows wild in extensive stands in desert and semidesert microclimates of northeastern Colombia and northwestern Venezuela, and it may be the most common member of the genus *Stenocereus,* except for *S. thurberi* of northwestern Mexico and Baja California, if domestic plants are included. Little is known of the distribution of *S. zopilotensis* or of its habits because it was only recently described, but it appears to be confined to the mid-Balsas region in Guerrero.

All pitayos are commonly planted in rows to produce living fences. In many indigenous villages, household yards are full of these species and others, many of which represent strains developed after centuries of selection for pulp color, fruit size, flavor, peel thickness, and relative number of spines. Between March and November, one sort of fruit or another can usually be found in fruit stands or hawked by street vendors in the native markets of Puebla and Oaxaca.

The Popoloca town of Reyes Metzontla, Puebla, in the far southwestern reaches of the Valle de Zapotitlán, is a center of columnar cacti diversity. At least thirteen species of columnars can be found within a 10 km radius of the village. The streets are lined and yards are crowded with the xoconochtlis and *pitayos de mayo*, plus chichipes *(Polaskia chichipe)*, chendes *(P. chende)*, garambullos *(Myrtillocactus geometrizans* and *P. schenckii)*, jiotillas *(Escontria chiotilla)*, and *malinches (Pachycereus marginatus)*. In adjacent hills and flats are thick forests of tetechos *(Neobuxbaumia tetetzo)*, with interspersed babosos *(Pachycereus hollianus)* and viejitas *(Pilosocereus chrysacanthus)*. Steeper hillsides feature giant specimens of órganos de cabeza amarilla *(Mitrocereus fulviceps)* and órganos de cabeza roja *(Neobuxbaumia macrocephala)*. Villagers demonstrate considerable expertise at harvesting the fruits and preparing them for consumption and marketing without incurring damage to themselves from the often potent spines. They are also inclined to carry on about the virtues of the different species.

The ripe pitayas (the pitayo fruits) usually grow high on the branches and will easily bruise if they fall to the ground. Careful collecting is essential because marketing times are critical, and any soft spot will tend to rot. In the highlands of central Mexico, pitaya harvesters use a curiously effective tool called a *chícole* for gathering the fruits. It is a long pole of an agave stalk tipped with a curious basketlike collector made of *carrizo*, probably *Arundo donax*, bamboolike cane that grows prolifically near water and in very moist soils. Several split canes are looped over the end of the pole and lashed, forming a series of wide loops spread evenly to form a porous basket. The ends form a closed, conical basket at the point where they are lashed to the pole. The collector nudges or wiggles a fruit loose with the side of a loop, and it slides or gently falls into the conical basket and can then be lowered and removed undamaged. If the fruits still retain spines (in most species the spines fall away when the

Figure 2.39. Harvesting the fruits of the pitayo de mayo (Stenocereus pruinosus), *San Juan Nochtixlán, Oaxaca. The orchard also contains xoconochtlis* (Stenocereus stellatus) *that yield fruit later in the season. The village name in Nahuatl means "Place of the Cacti."*

fruits are ripe), they are left for a few hours, and the spines that remain are easily scraped off. The same tool is used to gather other fruits, including chendes, chichipes, and jiotillas. Great care is taken so that the spines adhere to fruits intended for shipping to urban markets. Local harvesters note that leaving the spines attached retards ripening and allows for more successful marketing. The retailer removes all remaining spines at the point of sale. (Fig. 2.39.)

Literature from Mexican researchers concerning fruit production, commercial and personal, is abundant. Some government programs have encouraged the cultivation of these remarkable fruit producers, and it appears that cactus orchards are gradually replacing pastures weary of centuries of profound overgrazing by livestock. Cultivated orchards of various pitayas produce dramatically higher yields than wild stands. One study found the mean yield of cultivated stands of *S. pruinosus* and *S. stellatus* to be around 3000 kg (6600 lbs.) per hectare of fruit compared with less than 100 kg (220 lbs.) from wild stands. The plants yielded more and much larger fruits.[35] This statistic demonstrates the enormous economic potential of horticulture of columnar cacti. Some researchers suggest that the characteristics of cultivated varieties of xoconochtlis may be so distant from those of the wild varieties that the cactus is probably domesticated. If so, it will probably join *Stenocereus treleasei* and *Cereus hildmannianus,* the apple cactus, as the only columnars to have achieved that status.

S. pruinosus—called *cuepetla* or pitayo de mayo or sometimes *pitayo de octubre*—reaches more than 6 m in height (perhaps much more) but is often shorter. It is common in southern portions of Puebla, even more so in the central valleys of Oaxaca, where it is probably the most commonly cultivated columnar cactus. It has seven or fewer ribs (often the only easy way of distinguishing it from other pitayos). The fruits grow along the sides of the branch ribs and on the crowns as well. Vigorous specimens, perhaps wild, can be found growing on the hills above the valleys, but this cactus is even more common in towns and villages, where nearly every block features it among other cultivated cacti. In and around Santa María Acaquízapan, Oaxaca, near the state line with Puebla, peasants planted many hundreds of cuttings on overgrazed hillsides in 2001. They believe they will earn more income from the cactus fruits than they possibly can from grazing livestock. The village sits atop a mesa, and the orchards of several species, principally pitayo de mayo, have been planted on steep hillsides below. From late April through October, the orchards usually have one fruit or another ripe or ripening.

The fruits of this pitayo—the size of tangerines—begin to ripen in late April and continue through the first week of June. The fruits rot once the rains have begun (usually in mid-May), so the harvest tapers off rapidly after that. Harvesters gently wiggle the fruits free with chícole poles or, if they are within reach, twist them off by hand or with any old implement. Dead spines are left on the fruits as they are placed in packing crates to retard rotting. In the markets, vendors clean and sell them, usually for three pesos (about thirty cents U.S.). The fruits are of various shades of brilliant reds and purples, intensely sweet and satisfying. The outer husk is peeled off, and the entirety of the pulp is eaten. The seeds are tiny and can be chewed or simply swallowed whole. The syrup in the pulp has a tendency to run, so one must be careful to avoid dripping the deeply dyed liquid onto one's clothing. It is difficult to imagine a tastier fruit. I have eaten four at a sitting, but doubt if I could do that often. (Figs. 2.40, 2.41, and 2.42.)

Each plant yields from several dozen to several hundred fruits. Through the month of May, the fruits are common in markets, brought in fresh each day. Toward the end of the season, production dwindles to the extent that most fruits harvested are retained for local or household use. Some residents report that many of the plants flower once again in August and fruit in September into October, thus providing a second crop, a phenomenon unusual in columnar cacti of North America. This second harvest may be a result of centuries of selection and demonstrates the virtues of cultivation of this most esteemed plant. It also accounts for one of the plants' common Spanish names.

The taxonomy of *S. pruinosus* is troubling. Bravo-Hollis describes the plants as "from 4 to 5 m tall . . . ribs 5 to 6 [8]." Anderson supplies a similar description.[36] However, there are large numbers of cacti in lower elevations of western Puebla and southern Oaxaca answering to the description of *S. pruinosus* that exceed 7 m in height. Some extremely tall plants with six to seven ribs and fruits remarkably similar to those of *S. pruinosus* are to be found in coastal Oaxaca. They may reach 14 m in height. Several plants 10 m tall with six to seven ribs, hundreds of branches, and *pruinosus*-like fruits can be found growing sporadically on flat valley bottoms at 1500 m elevation south of Ejutla, Oaxaca. In the vicinity of La Reforma, at roughly 1000 m elevation midway between Oaxaca and Tehuantepec, plants around 8 m tall produce abundant pitayas that are sold in local markets. The plants are very common in the (apparent wild) vegetation of the region, often becoming the dominant large cactus in a region of great cactus diversity. These discrepant individuals may be a reflection of the lack of intense botanical studies in the region (doubtful) or, more likely, of the tendency for plants described in taxonomic literature to be those selected for fruit production in and around villages. These plants and their clones have long been selected for shorter habit, making access to the fruits easier come harvesting time. (Fig. 2.43.)

S. griseus, a cactus of tropical origins and possibly introduced into the southeastern Mexico from the Caribbean region, is common in the lowlands of Veracruz and northeastern Oaxaca, but less so (or not at all) in the central valleys. It also grows in huge numbers in the dryer zones of the Caribbean Rim, where plants may reach 10 m in height and grow in dense thickets. It is so common as nearly to be a weed in the desert scrub of Aruba, Bonaire, and Curaçao of the Netherlands Antilles, where it grows both wild and cultivated. It appears to grow wild in the mountain foothills of northern Veracruz, where individuals 9 m tall with dozens of branches are not unusual. I have not seen plants growing far from human habitation, however, and this pattern of distribution usually points to human introduction into the region. It is common in the dryer valleys and slopes of northern Colombia and Venezuela.

Figure 2.40. *Ripening fruits of pitayo de mayo, Acatepec, Puebla.*

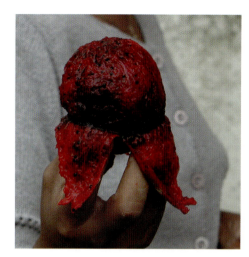

Figure 2.41. Pitaya de mayo, Oaxaca market.

Figure 2.42. Pitayas de mayo for sale, Huajuapan de León, Oaxaca.

*Figure 2.43. Pitaya (*Stenocereus *sp.), Oaxaca coast. This tree does not correspond to any published* Stenocereus *species.*

Figure 2.44. Stenocereus griseus, *Río Actopan, Veracruz. The plants appear both wild and in cultivation here and throughout the Caribbean region.*

Figure 2.45. *Crested yato* (Stenocereus griseus), *Aruba.*

The plants are also commonly planted around homes and gardens. They are to be looked for in gardens and along streets in the Río Balsas basin. *S. griseus* also is commonly horticulturally in lower valleys of southern Hidalgo and Querétaro, where I have found it growing in the Río Estorax valley in Querétaro, at around 1000 m elevation. (*S. pruinosus* appears to be absent below about 1400 m in that region, though a similar plant grows on the coast of Oaxaca and Guerrero.) Wild or adventive plants are common in the lush Río Actopan valley, east of Jalapa in northern Veracruz at elevations below 1000 m. The species seems to disappear above this altitude. Although I have looked for it in the Central Valley of Oaxaca, where several writers have pronounced it to be cultivated for fruit production, I have not yet been able to document a single plant.[37] I suspect that the 1500-plus m elevation of the valley is above its level of comfort, and its close relative, *S. pruinosus,* which seems to fare poorly at lower elevations where *S. griseus* thrives, replaces it. The two species appear to be closely related, which may account for the apparent confusion of *S. pruinosus* and *S. griseus* in some research. Until someone sorts out their distribution, confusion will continue to reign in ethnobotanical literatures. (Fig. 2.44.)

S. griseus plants are stellar fruit producers in the lower elevations of Mexico where they are common. They can be seen planted close to human habitation for easy access to the large, tasty fruits that come in a variety of bright colors in May and find their way into markets. Though these plants grow considerably taller then xoconochtlis (*S. stellatus),* up to 9 m tall, they may be difficult to distinguish from the latter when not in flower or fruit. The space between the areoles of *S. griseus* is characterized by a conspicuous hump, whereas in xoconochtlis the areole is located on the upper portion of the less conspicuous hump. In *S. griseus,* whitish flowers develop along the sides of the branches, not exclusively at the crowns, as is the case with xoconochtlis, and the *S. griseus* fruits are larger. The two species are seldom confused, however, for their distributions do not often overlap.

Millions of *S. griseus* plants fill many landscapes of the Netherlands Antilles, where they are called *yatos.* On Bonaire, Washington-Slagbaai National Park is home to legions of plants of great age. If they are adventive (introduced from outside and spreading afterward), they were introduced on the islands long ago, for the large plants appear to be well established on the island's wildest places. (Fig. 2.45.)

Figure 2.46. Yato fence, Bonaire.

Yatos appear to thrive on the arid limestone and fossil coral (and occasionally volcanic) soils of these small islands. In scattered areas, they might legitimately be considered weeds, so great is their propensity to sprout up, especially on soils disturbed by goats. Although the fruits are widely eaten, there appears to have been little selection for fruit production, and one seldom sees plants heavy with fruits, as is the case in Mexico. In spite of the sweet flavor with a subtle spicy taste, natives seem to view the fruits with indifference, preferring those of the kadushi *(Cereus repandus),* which are free of spines. Most of the yato fruits have white pulp, although red pulp and yellow pulp are apparently not uncommon. This characteristic of the Netherlands Antilles plants also contrasts with those in Mexico, where red to purple pulp is the norm for *S. griseus,* and white is uncommon. In the Netherlands Antilles (and perhaps elsewhere), *S. griseus* flowers more than once each year, possibly repeatedly.

Although yato fruits are well known in the Antilles, by far the most common use of *S. griseus* is in the construction of living fences. They appear throughout the rural parts of the islands of Aruba, Bonaire, and Curaçao. In the granitic uplands of Aruba, the fences proliferate in mind-boggling numbers and sizes and create a most agreeable patchwork of lots. The yato is surely the most common large plant on the island, and virtually no house is without its share, each usually sporting a yato fence and a couple of large plants as well. On Bonaire, the rapidity of growth of yato fences has given rise to a fence industry and a peculiar configuration that lends itself to a most satisfying design. Inexpensive rabbit wire with mesh roughly 15 cm wide is strung at 1 m above the ground. Lopped-off yato arms are woven through openings in the wire, each new arm directed to the opposite side of the previous branch, thus forming a crude X. The scarred portion of the cut branch is left in contact with the ground and sprouts almost immediately. The branches may grow more than 30 cm a year. The resulting fence is burro and goat proof and provides superb privacy. During the rainy season, there is a tendency for climber vines (especially *Antigonon leptopus*) to add to the apparent solidity of the now wall-like fence. There is no need to water except in times of extreme drought, so once the fence is complete, it is virtually maintenance free. When the rabbit wire rusts away, its absence is not noticed because the yatos have assumed greater strength than the wire could ever afford. Thus, tourist facilities and hotels have taken to planting yato fences with the secure knowledge that in a couple of years' time the fence will be as secure as virtually any wall (though not machete proof!). (Fig. 2.46.)

Figure 2.47. Xoconochtli (Stenocereus stellatus), *Río Acatlán valley, Puebla.*

Xoconochtli (*S. stellatus,* pronounced *show ko NOACH tli*) fruits ripen late July through August. The delicate pink flowers open at night in May and June, when the pitayos de mayo *(S. pruinosus)* are already in fruit, and are pollinated primarily by nectar-feeding bats. The plants are distinguishable from other common pitayos by the position of the flowers and fruits around the tips of the branches like a crown. Xoconochtlis are also distinguished from pitayos de mayo by their eight to twelve ribs on the branches (as is also the case with *S. griseus*), whereas pitayos have seven or fewer. The areoles are raised on the ridges of the branches and are covered with stout spines, producing a knobby effect. (Fig. 2.47.)

Villagers usually pronounce fruits of xoconochtlis to be every bit as good as the pitayas de mayo. They ripen at a time when few other cacti are producing fruits. They are somewhat smaller than other pitayas and often (but by no means always) have a slightly piquant taste. The pulp ripens in a variety of colors—yellows, purples, reds, and even white. Villagers consider red pulp to be the wild color. I suspect, however, that white pulp occurs in wild specimens as well, as I have found to be the case with *S. thurberi* and

Pachycereus pecten-aboriginum. Yellow pulp is more common in *S. stellatus* than in any other *Stenocereus* species I have seen. Casas, Caballero, and Valiente-Banuet note the different uses of fruits with different-colored pulp:

> people generally consider the best xoconochtli to be those with large fruits, white sweet pulp, thin peel and few spines. However, they continue to maintain variants with different characteristics because they are used for different purposes. For instance, sour fruits are good for preparation of drinks and jams; spiny fruits are more resistant to predators; fruits with thick pericarp resist rotting and are better for long distance transportation. Artificial selection thus seems to favor a number of different combinations of fruit characteristics, each playing a particular economic or cultural role.[38]

Residents identify and distinguish xoconochtli fruits as readily as Europeans living in apple or cherry country know those fruits. Indeed, the fruits of the various cacti are as taxonomically different as are tree fruits of the rose

Figure 2.48. Xoconochtli fruits for sale in market, Acatlán, Puebla.

Figure 2.49. Xoconochtli buds and flowers, Reyes Metzontla, Puebla. Note that they are confined to the tips of the branches.

family in northern climates—apples, apricots, peaches, and pears—and the varieties among species rivals the varieties of temperate fruits. In spite of the variation in cactus fruits, villagers in Puebla and Oaxaca were reluctant to express any hint of favoritism when I asked them which fruits they liked the best. I think I understand. After all, which is better, an apple or a peach? (Fig. 2.48.)

In central Mexico (unlike in northwestern Mexico), when fields are cleared for planting (or recleared after a period of fallow), columnar cacti are usually spared. They are recognized as a valuable resource and afforded a degree of protection generally reserved for orchard trees. Thus, it is often difficult to determine whether a given plant is wild, cultivated, or merely one that has been long managed in situ.

In addition to the consumption and marketing of the raw fruit, an indigenous alcoholic beverage called *colonche* is brewed from the cactus fruits in the region. I have not tried it. The seeds, when toasted, are also rather tasty and popular. The flesh of new growth—tips of branches—was formerly a well-known dish, but that culinary custom resulted in elimination of fruits from the lopped-off branch. Many people still consider flowers added to soups or fried with eggs a delicacy. In times of drought and frequently in the hot months of March and April, peasants will lop off branches, singe the spines, and feed the flesh to livestock. Goats consume buds, flowers, and fruits that may fall to the ground. The wood of dead plants is an important source of fuel for firing pottery and in the kitchen. Xoconochtli cuttings (and those of pitayo de mayo as well) constitute an important plant for erosion control. The central valleys and uplands of Oaxaca, especially the Mixteca Alta,

Figure 2.50. Xoconochtli fence, Oaxaca. The fence keeps livestock in or out and produces fruit.

exhibit frightful erosion, and seeing these useful plants doing double duty as fruit producers and soil stabilizers is heartening, indeed. (Figs. 2.49 and 2.50.)

Figure 2.51. Tunillo buds (Stenocereus treleasei), *Oaxaca. Note the similarity to xoconochtli buds, fig. 2.49.*

Figure 2.52. Tunillos, Oaxaca. No wild plants of this species are known.

In short, xoconochtli is just about as useful as a plant can be.

S. treleasei, the *tunillo,* is a smaller pitayo, rather common in the vicinity of the city of Oaxaca, less so in other parts of the central valley of Oaxaca. Anderson notes that the tunillo is uncommon, but I found it to be abundant within a 30-km radius of the city of Oaxaca, even growing at the Oaxaca airport and around the Zapotecan archaeological site of Monte Albán, where it is planted as an ornamental. It is usually a single stalk or several single stalks emerging from the ground and has up to twenty ribs, far more than any other *Stenocereus* of the region. The reddish fruits are edible, but due to the relative scarceness of the plant are not marketed widely, if at all. Tunillos can be seen scattered about pastures and on gentle slopes on the periphery of ancient fields. Cornejo reports that in twelve years of field-work in and around Oaxaca, he never saw an unequivocally *wild S. treleasei.*[39] Nor have I. (Figs. 2.51 and 2.52.)

Cornejo notes that *S. treleasei* has roughly double the number of ribs of *S. stellatus.* Floral buds and flowers are arranged in both species at the apex of the ribs. He hypothesizes that increasing the number of ribs increases the number of areole sites and thus the number of potential fruits, so *S. treleasei* may be a cultivar derived from a variant cutting of *S. stellatus.*[40] Gibson suggests that apart from the different number of ribs, the two species have little to distinguish them and perhaps should be considered one species.[41] Given the apparent absence of wild individuals of *S. treleasei,* it has seemed prudent to some to view it as simply a domesticated variant of *S. stellatus.* We can even imagine a peasant puzzling over an odd-looking xoconochtli, one with an abnormally high number of ribs, lopping off a couple of arms, carrying them home (in a bag woven of agave fiber!), and planting them in the family plot. Tunillos, it seems probable, are a horticultural derivative of xoconochtlis. Because they do not appear to interbreed or commonly hybridize with xoconochtlis, they may be a species that has evolved due to human manipulation. Human intervention in evolution is clearly seen in the ancient varieties of columnar cacti of southern Mexico. Here we may have a new species, not just a new variety.

Three additional species of the genus merit discussion here.

Figure 2.53. Stenocereus chacalapensis, *Huatulco, Oaxaca. The largest* Stenocereus, *these plants grow only within a radius of 30 km (17 mi.).*

Figure 2.54. Stenocereus chacalapensis *trunk, Huatulco, Oaxaca.*

Stenocereus chacalapensis (**Bravo et MacDougall**) **Buxbaum**
pitayal

This species does not grow in the central valley of Oaxaca at all. It appears to be confined to a relatively small area of rich (and rapidly disappearing) tropical deciduous forest along Oaxaca's southern coastline. The plants have thick trunks that resemble those of coniferous trees and reach at least 15 m in height, quite possibly more—that's 50 ft. or higher. Indeed, *S. chacalapensis* rivals *Pachycereus weberi* and *P. pringlei* in size and may even exceed them in height. The branches often grow rather closely together, giving the plant a compact, perhaps bottom-heavy appearance and making its great height difficult to estimate. The giants are also difficult to photograph in their habitat due to the interference of the intertwined canopy of other trees. The species is named after the type locality, the village of San Isidro Chacalapa, Oaxaca (not the village of Chacalapa a few kilometers inland

from the coast on the grueling Oaxaca-Huatulco highway!). I found several of the plants along a 2-km stretch growing in thick tropical deciduous forest along the undulating highway a few kilometers east of the resort district Bahías de Huatulco, but none thereafter. The husks of the fruits are bright red and similar in size to other pitayas, quite attractive in their brilliance, but protected by a thick covering of yellowish spines. The spines are distinct in that they fall from the fruits as a unit and reach the ground in the shape of a basket (as do those of a similar species, *S. chrysocarpus* of southern Michoacán, and those of *Armatocereus matucanensis* in Ecuador). The pulp is also bright red. I have been unable to taste the fruits due to their inaccessibility on the lofty branches. If the number of birds devouring the fruits is a test of their succulence, they are tasty, indeed, for birds seem to become fearless when devouring the pulp. Natives do not appear to distinguish between this species and a similar but distinct (and undescribed) species of *Stenocereus* noted later. (Figs. 2.53 and 2.54.).

Stenocereus laevigatus (Salm-Dyck) Buxbaum
pitayo

This rather nondescript pitayo is found in the more temperate valleys of southern Chiapas and quite probably in southeastern Oaxaca as well. I am uneasy as to the correct taxonomy of the plants. As described and photographed by Bravo-Hollis,[42] they have seven ribs, not the eight to ten stated by Anderson,[43] and have one central spine 2.5 to 3 cm long on each areole. The large plant in figures 2.56a and 2.56b conforms to number of ribs and spine arrangement set out by Bravo-Hollis. The areoles are indistinct, in clear contrast to felty areoles of the most probable alternative, *S. pruinosus,* which has also been reported from Chiapas. The spines are also somewhat narrower than those of the latter. The range of variation of characteristics of the *Stenocerei* in this region is so great that I suspect these two species may hybridize, and the union may explain the confusing taxonomy. In the vicinity of the Isthmus of Tehuantepec, I have seen *Stenocereus* species with four, five, six, and seven ribs, all with varying arrangements of spines that do not conform to any published descriptions. Some plants have especially stout and long spines. I have photographed similar cacti in the Yucatán. Their description does not match that of any *Stenocereus* herein reported, but is probably *S. eichlamii.* (Fig. 2.55.)

I found a large plant of what appears to be *S. laevigatus* growing almost exactly where Bravo-Hollis noted, just east of the city of Cintalapa, Chiapas. It was a solitary individual about 8 m tall, left to sunburn and die when the slope on which it grows was cleared for pasture. Other pastures featured similar plants. I suspect that they grow only on hillsides, for I searched in vain in the flat tropical deciduous forest west of the city. Another plant flourished on steep volcanic substrate along the roadside in the Reserva de la Sepultura on the Oaxaca-Chiapas boundary. Smaller plants have sprouted on roadway cuts in this rich tropical semideciduous forest, probably unable to compete with the luxuriant vegetation growing on the uncleared slopes. (Figs. 2.56a and 2.56b.)

Natives of the region appear to pay little heed to uses of this or other cacti, an understandable oversight given the abundance of cultivated and spineless orchard fruits.

Stenocereus eichlamii Britton & Rose Buxbaum
pitayo

This pitayo of the Yucatán Peninsula is similar to both *S. laevigatus* and *S. pruinosus,* but differs in rib count and spination. It may eventually prove to be the same species as one or the other. It is a common plant in the thornscrub of the northern portion of the peninsula. (Fig. 2.57.)

Figure 2.55. Unknown Stenocereus, *long spines, few ribs, Salina Cruz, Isthmus of Tehuantepec.*

Stenocereus zopilotensis Arreola-Nava & Terrazas

While I was traveling in the Balsas basin in 2002, I noticed a tall cactus that resembled a xoconochtli growing on a steep hillside of a side valley of the Cañon de Zopilote. I photographed it, nearly breaking my neck to do so. The fruits grew near the apex of the branches, but were not positioned on top, as is usually the case with *S. stellatus.* I came upon a second plant not far away. In labeling the photo, I found no reference in any literature to a different species, so in my haste I labeled the photograph "xoconochtli from Zopilote" and hurried on. To my surprise, more than two years later I happened upon a monograph in which a new species of *Stenocereus* from Cañon de Zopilote was described.[44] It fit the description of the cactus in my photographs perfectly. I have no information on the ethnobotanical uses, but the fruits are surely sweet and juicy. (Fig. 2.58.)

This species is reported only from Cañon de Zopilote and environs. I suspect that it will be found to be more widespread as further explorations produce more information.

Figure 2.56a. Stenocereus laevigatus, *Cintalapa, Chiapas. This plant is healthy even though the forest has been cleared around it.*

Figure 2.56b. Close-up of Stenocereus laevigatus. Note the long central spine that helps identify the species.

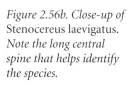

Figure 2.57. Stenocereus eichlamii *in thornscrub, Dzibilchaltún, Yucatán.*

Figure 2.58. Stenocereus zopilotensis. *This newly described species appears to be confined to the environs of Cañon de Zopilote of the Río Balsas in Guerrero.*

Stenocereus sp.
pitayo

This unidentified and probably undescribed species is uncommon but remarkable indeed, growing in a restricted area in the southern portion of Oaxaca's central valley. I have found a similar plant growing along the Costa Grande of Guerrero and along Oaxaca's coast as well. Locally called *pitayito,* it is a huge plant, reaching nearly 12 m in height, possibly more, with many dozens of branches forming a dense habit. The fruits are similar to those of *S. pruinosus,* but smaller. These giants, with branches growing in many directions, appear as solitary individuals in the fields, where due to their massive size they can be seen from afar. I would like to have taken more photographs of this superb species, but a resident swarm of angry bees took exception to my presence in the vicinity. (Fig. 2.59.)

Perhaps the same species, but probably a different one, judging from the orientation and organization of its branches, occurs in tropical deciduous forest on the south coast of Oaxaca. Neither its description nor its habitat corresponds to any species that has come to my attention, which seems odd indeed, given the great size of the plant and its conspicuous location. In 1970, Mexican cactologist Hernán Sánchez-Mejorada described finding a similar columnar on the coast of Jalisco: "[It has a] . . . thick trunk that begins to branch at a height of 3 to 5 m above the ground, sometimes more, whose large arms, of five or six ribs, measure 10 to 16 m long. This colossus of the forest appears to resemble *Ritterocereus* [*Stenocereus*] *chacalapensis,* a native of Oaxaca, but should be carefully studied, for we were not familiar with the flowers and fruits of the . . . specimens, apparently quite different from the fruits of *R.* [*S.*] *chacalapensis* in that they are not covered with as many spines."[45]

Near Crucecita and San Isidro Chacalapa, this unnamed cactus grows to at least 15 m in height. The red pulp of the fruits is sweet. Harvesting is difficult, however, due to the plants' great height and the necessity of using long poles to dislodge the fruits. The plants are similar to etchos in habit and in height. The areoles become conspicuously knobby, far more so than those of *S. pruinosus.* The highest branches appear to reach the canopy of the local forest, which perhaps yields a clue to their great height. (Figs. 2.60 and 2.61.)

Figure 2.59. Undescribed species of Stenocereus, *different from the ones shown in figs. 2.60 and 2.61, near Ejutla, Valle Central de Oaxaca. Note also the difference in habit.*

Figure 2.60. Undescribed species of Stenocereus *near Pochutla, coast of Oaxaca.*

Figure 2.61. Undescribed species of Stenocereus *near Huatulco, coast of Oaxaca, different from the one shown in fig. 2.60. Note the long trunk and close habit of the branches.*

We may be dealing here with two undescribed species, one found in the central valleys, the other along the coast. Or they may simply be variants of a described species, most likely *S. laevigatus* or *S. pruinosus*, which they resemble in their fruits and in their rib number. (Figs. 2.62 and 2.63.)

Pitayos of the North

Stenocereus martinezii (J. G. Ortega) Buxbaum 1961

Stenocereus martinezii is rather poorly known in comparison with *S. thurberi*. Its range is considerably smaller than that of the organ pipe, hardly more than 100 km along the Sinaloan lowlands of western Mexico. Its noticeable trunk is a giveaway that it is a tree of the forest, not of the desert or thornscrub, even if its height seldom exceeds 5 m.

Stenocereus gummosus (K. Brandegee) A. C. Gibson & K. E. Horak
pitaya agria, *ziix is ccapxl* **(Seri), galloping cactus**

S. alamosensis (J. M. Coult.) Gibson & Horak
sina, galloping cactus

S. kerberi (K. Schum.) A. C. Gibson & K. E. Horak

These three cacti are only marginally columnar, tending more toward shrubbiness. They tend to grow in sprawling thickets. At times, however, individual branches take on the uprightness of columnars that exceed 2 m in height; hence, I include them in this study. All three species grow near the Gulf of California.

Stenocereus gummosus, pitaya agria, is the only one of the three found in Baja California, where it is quite common in the Sonoran Desert portions of the peninsula and on Tiburón Island (part of Sonora). It grows in dense thickets whose tallest branches reach little more than 3 m tall. On the Pacific Coast, plants can be found well to the north, nearly to Ensenada, but these plants tend to be shrubbier, often seeming to grow horizontally under the influence of the winds. Natives report that the Pacific-region plants also produce fewer fruits. (Fig. 2.64.)

Figure 2.62. Undescribed species of Stenocereus, *near Huatulco, Oaxaca.*

Figure 2.63. Pitayas, perhaps from undescribed species, coast of Oaxaca.

Figure 2.64. Pitaya agria (Stenocereus gummosus), *Baja California. Many people consider this species' fruits the best of all cactus fruit.*

*Figure 2.65. Flower of
Stenocereus gummosus,
Baja California.*

On the Sonoran mainland, the range of the pitaya agria, as it is widely known, is restricted to the immediate vicinity of Punta Chueca and El Desemboque, the two present Seri villages, and of several seasonal camps where until the final decades of the twentieth century Seris dwelled while following changing fishing conditions. A large plant growing on a stabilized dune above the beach some 2 km south of the fishing village of Bahía Kino is apparently the southernmost mainland plant, and another growing at the southern end of the village of El Desemboque is the northernmost. This immediacy to Seri habitations suggests that the plant, which is an important fall food source for the Seris, grows on the Sonoran mainland because the Seris brought it there from Baja California, as Alberto Búrquez and I have argued.[46] It is far more common on Tiburón Island (where Seris had numerous camps), but even there it tends to grow primarily along ancient Seri trails or near places where Seris would congregate. For the Seri name, I have used the transliteration employed by Felger and Moser, *ziix is ccapxl,*[47] roughly pronounced *sheej-ees-ccapl.* They explain that the Seris changed the name of the plant when a Seri child, whose name resembled the old name for the cactus, died. Seri custom required that if a person bore the name of a plant or animal, at the person's death the animal or plant name was changed.

The Seris chose the plant well. Under optimum conditions, *gummosus* fruits are the largest and probably the tastiest of all the wild fruits found in northwest Mexico, comparing in size and flavor with the horticulturally selected fruits of Oaxaca and Puebla, but with a more piquant taste. This characteristic should not be surprising, for Gibson and his associates, using chemical analysis, determined that *S. gummosus* and the prolific fruit producer of Oaxaca, *S. stellatus,* the xoconochtli, are closely related.[48] Furthermore, the plants' growing season is adapted to the seasonality of Baja California, whose skimpy rains fall primarily in late summer, fall, and winter, rather than in midsummer. Flowering occurs during the summer and early fall months. The white to pink flowers are apparently moth pollinated and bloom from July through November. The flower illustrated in figure 2.65 was blooming in early November. The fruits ripen in September well into November and thus become available to humans when all other cactus fruiting seasons have finished. Baja California plants are widespread and abundant and seem to thrive with annual rainfall of 100 mm or less—that's only around 4 in. (Fig. 2.65.)

Harvesting the baseball-size fruits is not a simple matter, for they are usually well beyond human reach, embedded enticingly within the fortresslike confines of a mishmash of arms, each areole of which contains a daggerlike central spine that may exceed 5 cm in length. The Seris long ago perfected harvesting, however, so if one is fortunate enough to have Seris to assist, harvesting pitayas agrias should be no problem.

I have several times watched them harvest the fruit. They fashion a pole, often from a saguaro or sahueso rib or two lashed together, then fashion a point from whatever wood is available, and attach the point to the pole. With measured dexterity, they select the most advantageous approach, spear the fruit (when the fruits are ripe, the tough spines on the husk of most plants tend to fall away) and carefully wiggle it from the stem. The husk is rather easily removed, and the pulp is a gourmet delight. I found the fruits to have a flavor reminiscent of raspberries. The Seri country is very hot well into the fruiting season of the pitaya agria, and the fruits are not only tasty but most effective thirst quenchers as well.

The Seris were until the past three decades nomadic hunters and gatherers and, apart from planting useful desert plants in the vicinity of their temporary camps, were not known to practice permanent agriculture. Hence, they have no recorded history of selecting, planting, and nurturing the most productive plants, but over their eons of living in the desert they may have manipulated the pitaya agria to their purposes. It seems to be an ideal candidate for cultivation. In its natural setting, it routinely flourishes with annual rainfall of less then 100 mm. A program to establish experimental orchards of pitaya agria could be carried out at minimum cost and with only marginal inputs. The Seris know the most prolific fruit-producing plants and might easily dedicate tiny portions of their lands to raising this most worthy fruit. If, as Gibson suspects, pitaya agria is closely related to the xoconochtli, *S. stellatus,* the prolific fruit producer of Oaxaca and Puebla, selection for fruit production should be beneficial indeed. Israeli scientists have recently begun experiments planting pitayas agrias in the Negev Desert.

Stenocereus alamosensis, the sina, is well known to coastal and valley indigenous peoples of Sonora, especially Mayos and Yaquis. The plant grows abundantly in coastal and foothills thornscrub, but disappears where the thornscrub merges into the Sonoran Desert. It is found in rambling thickets up to 10 m in diameter, similar to pitaya agria, but the stems or branches are thinner, and they grow less densely. Even so, the thickets are intimidating, and many of the tasty fruits go unharvested due to lack of easy access. The sina fruits also tend to ripen simultaneously with pitayas, and in terms of taste and ease of gathering they come out second best. I have seen (although rarely) stems nearly 4 m in height, so a few plants qualify for columnarhood, but usually they are only half that tall and flunk the test.

Sinas are the only member of their genus with red (grading to shocking pink), tubular diurnal flowers, apparently pollinated by hummingbirds. The flowers also grow well down the branches from the apex, quite different from most other members of the genus, which tend to flower at or near the top of the branches. The stems of stressed or older plants are often reddish below the new growth, a

Figure 2.66. *Sina fruit* (Stenocereus alamosensis), *Sonora.*

phenomenon I have not seen in other cacti. The color makes the plant look sickly, but appears in no way to interfere with its growing, flowering, and fruiting.

The red fruits are the size of ping-pong balls and ripen throughout the summer. They are tasty, often drier than fruits of the pitaya (*S. thurberi*) and not as sweet. Plants produce fewer fruits than is the case with pitayas. Sina fruits usually have few spines or none at all, but the large expanse of spiny cactus stems sometimes renders the ripened fruits accessible only to animals, which usually harvest the fruits before humans can gather them. (Fig. 2.66.)

These fruits, in addition to their fine taste, are also reported to help ease the pains of rheumatism and to alleviate the pain and swelling of insect stings. A Mayo friend of mine collected the husks of ripe fruit to boil into a tea for soothing his granddaughter's heat rash. Mayo weavers, who produce densely woven woolen blankets, use a length of cleaned and polished stem wood to elongate the *na'abuia,* or shed shuttle, of their *telares* (ground looms) to accommodate wider blankets. The wood is carved into a cylinder with a tapered end and inserted into the hollow end of the stiff carrizo grass (*Arundo donax,* Poaceae) that forms the shed shuttle. The width of the blanket can thus be extended indefinitely. The sina wood is strong enough to withstand the constant tamping pressure of the *sasapayeca* (tamping stick made of etcho wood) and the moving warp.

S. kerberi is similar in growth habit and appearance to the sina. It grows in Sinaloa, where it seems to flourish in tropical deciduous forest. Its flowers are pink rather than red, and the branches have four ribs, as opposed to the usual seven for sinas. I have not seen it, but based on photographs I have viewed, I do not consider it a columnar cactus.

The True Giants: The Genus *Pachycereus*

Included in the genus *Pachycereus* are the two most massive species and perhaps the tallest as well. Unlike the genus *Stenocereus,* these species exhibit a dramatic variation in appearance of their stems, branches, flowers, and fruits. Even a sophisticated cactologist would be surprised at first glance to learn that *Pachycereus hollianus, P. marginatus,* and *P. pringlei* are members of the same genus. *Pachycereus* is nearly a purely Mexican genus. One species *(P. schottii)* makes it into a small corner of extreme southwestern Arizona. Another *(P. lepidanthus)* is found in Guatemala.

Pachycereus pringlei (S. Watson) Britton & Rose
sahueso, cardón, *xaasj* (Seri)

I begin this splendid genus with the sometimes monstrously large sahueso (its Sonoran name), or *cardón sahueso.* It grows along the very dry Gulf Coast region of Sonora and in much of Baja California, where it is called simply the *cardón,* a name it shares with its cousin *Pachycereus pecten-aboriginum.* It, the chico (*Pachycereus weberi*), and *P. grandis* from southern Mexico are the top candidates for the title of the Biggest Cactus. Oddly enough, the biggest sahuesos seem to grow where the rainfall is nearly the lowest. Among the most massive individuals are those in a grove near Bahía Kino, Sonora, where annual precipitation is less than 90 mm. Other colossi appear in even drier terrain near Bahía de los Angeles, Baja California. Great groves of sahuesos appear in unlikely places—on highly arid small offshore islands and on relict sand dunes along the Sonoran coast—as well as in very dry portions of Baja California and on all the islands in the deeper Gulf of California, often on rocky slopes. Large sahuesos weigh up to 25 tons.[49] In their ability to grow to elephantine size in a region of markedly limited rainfall (in parts of Baja California and Sonora a year may pass with no measurable precipitation), sahuesos are a miracle of evolution. Among large columnar cacti, only the soberbio *(Browningia candelaris)* of Chile and Peru, a far smaller plant, and plants of the genus *Eulychnia* of the Atacama Desert and *Neoraimondia arequipensis* of Peru can flourish in such extreme aridity. (Fig. 2.67.)

The sahueso's ability to withstand prolonged droughts has puzzled visitors for centuries. Fr. Miguel del Barco, who spent considerable time in missions of Baja California near the middle of the eighteenth century, made the following observations:

> Whence come that moisture, and that liquor, of which it is so full? Not from the rains, as these are very scarce in California, and it is for this reason that,

Figure 2.67. *Grove of large sahuesos* (Pachycereus pringlei), *Kino Bay, Sonora.*

> unless there is a permanent water source and of sufficient size that watering can be done with adequate frequency, nothing can be sown or planted, trusting only to rain water, as it will dry up in a short time. The cardón, even though years may pass without rain, does not show any feeling on account of this, but it perseveres with the same serenity, with the same green and fresh color, and with the same moisture as always. . . . [It] has so much juice inside that there is no European tree that can equal it in this respect, and even perhaps in America itself there is no tree which surpasses it in this property.[50]

Sahuesos are possibly the tallest columnars. Thousands of the huge plants exceed 12 m in height, more than 40 ft. Salak measured and photographed an individual 19 m tall in Baja California, where the tallest trees grow.[51] Such giants often prosper in the protection of nearly inaccessible canyons, so taller specimens may yet be found. Exceptionally tall populations are found among granitic boulders

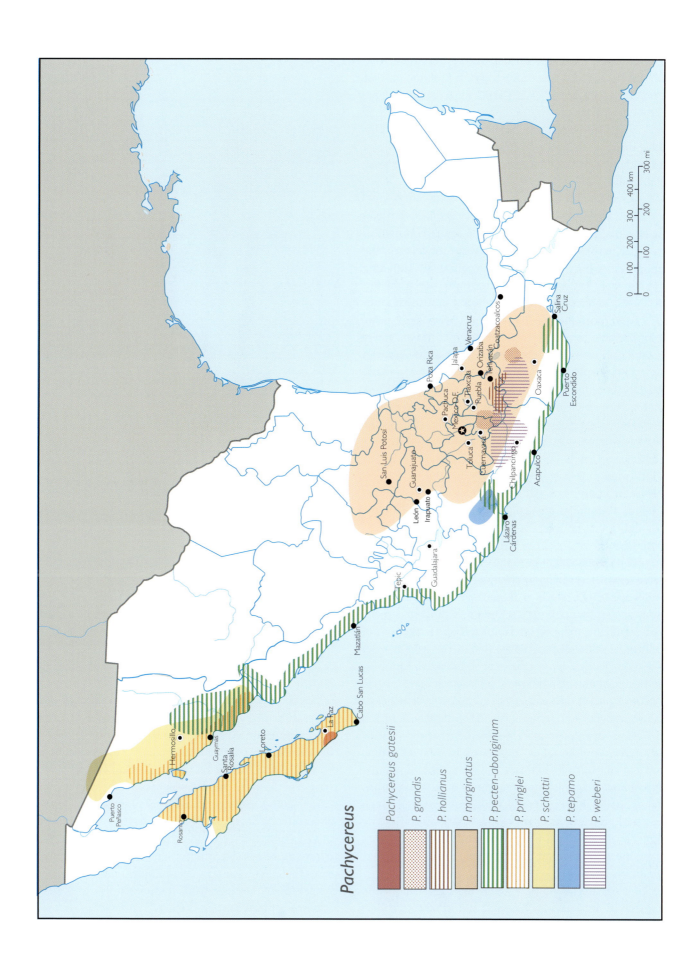

Pachycereus

Pachycereus gatesii
P. grandis
P. hollianus
P. marginatus
P. pecten-aboriginum
P. pringlei
P. schottii
P. tepamo
P. weberi

Puerto Peñasco
Rosario
Hermosillo
Guaymas
Santa Rosalía
Loreto
La Paz
Cabo San Lucas
Mazatlán
Tepic
Guadalajara
San Luis Potosí
León
Irapuato
Guanajuato
Pachuca
Poza Rica
Jalapa
Veracruz
Orizaba
Tlaxcala
Puebla
México D.F.
Toluca
Cuernavaca
Tehuacán
Coatzacoalcos
Oaxaca
Salina Cruz
Puerto Escondido
Chilpancingo
Acapulco
Lázaro Cárdenas

300 mi
400 km
300
200
100
200
100
0
0

near Cataviña, in midpeninsula, and in granitic arroyo soils on the western side of the Sierra la Libertad, not far from the Sea of Cortés. Where the sahueso cohabits in large numbers with the bizarre boojum tree *(cirio, Fouquieria columnaris)*, the landscape becomes eerily otherworldly. Other massive sahuesos grow near La Paz, in Baja California Sur, and on Santa Catalina Island in the Gulf of California. (Fig. 2.68.)

I suspect that the sahueso is the most abundant columnar cactus in nature, with the *Stenocereus thurberi*, organ pipe cactus, a close second, and the saguaro third. All three are natives of the Sonoran Desert. In Sonora, the sahueso is confined mostly to a strip perhaps 300 km long and barely 40 km wide (usually narrower) along the Sea of Cortés, whereas the saguaro is found in a continuous wide band more than 1000 km long. The sahueso grows in impressive numbers and over a much larger range in Baja California, however. In portions of its Sonoran range in close proximity to the sea, it may occur in dense semiforests of massive individuals. South of Bahía Kino, Sonora, thousands of mature sahuesos populate stabilized Pleistocene sand dunes parallel to the shore, often in the lee of active dunes. These individuals appear to be part of a relict stand, for they are mostly of uniform age, and there is little recruitment of seedlings. Whether this outcome is due to the ubiquity of wild and domestic herbivores that nibble or trample young plants or to an historical climatic singularity that favored a single recruitment (or series of recruitments) will require further research. The great plants may have germinated prior to the introduction of livestock in the region. Elsewhere, especially where the plants receive protection from herbivores, recruitment is successful, and many young plants flourish. On small protected rocky islands off Guaymas and off Tiburón Island in the Gulf of California, sahuesos in a remarkable mix of ages grow in truly unbelievable density that must be seen to be appreciated—as many as eight thousand plants per hectare.[52] These forests of cacti greatly exceed the great cactus densities of *Stenocereus thurberi* in the Pitayal in southern Sonora and of *Neobuxbaumia tetetzo* in the tetecheras of the Valle de Tehuacán in southern Puebla, Mexico. Isla Santa Catalina north of La Paz, Baja California, has a magnificent population of huge plants. Throughout much of Baja California, stands are less dense, but one can drive for hours through scattered populations of the cactus that seem to have no end. The sahueso is a truly successful drought-tolerant giant, the world's largest plant growing under such desert conditions.

Sahuesos are sympatric with etchos in a small area in the vicinity of Guaymas, Sonora, and in southern Baja

Figure 2.68. Sahueso, Cataviña, Baja California.

California, where distinguishing between the two often proves difficult except during flowering and fruiting time. I once engaged in a lively debate with a prominent botanist over the identity of several large plants growing well south of Loreto, Baja California Sur, on the Gulf of California coast. The friendly dispute was never settled. In general, etcho fruits have several times as many long spines, whereas sahueso fruits have fewer, barely extending beyond a velvety covering that resembles cattail fluff. Sahuesos tend to have more massive trunks than etchos, and on smaller plants their areoles are far less prominent than those of the etcho. Pulp of the etcho fruits is more reddish, and that of sahuesos is purplish. (Fig. 2.69.)

Figure 2.69. Sahueso fruits. The pulp is more purplish than red. Note the large seeds.

In the Gulf Coast region of Sonora, sahuesos often intermingle with saguaros, though on a limited basis. In the absence of flowers and fruits, they can usually be distinguished by the difference in rib numbers (sahuesos have ten to fifteen, saguaros twelve to twenty-five) and by the saguaros' tendency to grow as a main stalk from which arms or branches emerge and seldom exceed the height of the trunk. Sahueso branches often exceed the height of the vertical trunk axis, and it is often difficult to perceive a main stalk. Sahuesos usually have a more succulent appearance. They are more massive and taller on the average than saguaros, and with decent rainfall they grow faster. A plant I knew well grew in ten years from a stubby single-trunked meter-high juvenile to a massive three-armed sentinel 5 m tall. It apparently profited smartly from its location over a septic tank. Saguaros also respond well to supplemental watering, but none appears to grow at anywhere nearly the speed of sahuesos.

Where sahuesos are abundant in Sonora, saguaros are usually either uncommon or absent, in part due to sahuesos' affinity for a variety of soils—rocky, sandy, or clayey and poorly drained. Saguaros tend to prefer well-drained, coarse, or rocky soils. Sahuesos also appear to be considerably more drought tolerant than saguaros and thus abound in habitats too dry for exuberant proliferation of saguaros (which is probably why no saguaros grow in Baja California). However, sahuesos are not frost tolerant, whereas saguaros appear to do fine with a few degrees of frost.

The sahueso's root system is a marvel to behold. In sandy or loamy soils, roots radiate from the trunk in a huge network not far below the soil surface. I tracked one

such root 15 m from the base of the plant. In another case, a bark-covered root nearly 10 cm thick emerged from the soil more than 10 m from its source, stayed above the surface for a meter, then dived into the sandy loam, only to surface again a few meters farther away. The roots of this vast network function as outriggers to anchor the giants and prevent them from being blown over, especially from the fall hurricanes that from time to time lash the region. Furthermore, the location of roots at or near the surface not only lets them intercept the scantiest of rainfalls, but also harvest the heavy dews characteristic of the gulf. In the great cardonales (as sahueso forests are called), the network of individual roots produces such a multitude of far-reaching tentacles that they form one vast interwoven mat of mutually thieving, but also mutually supporting roots. (See fig. 3.8.)

On the steep rocky slopes of Midriff Gulf of California islands, almost no soil is available for the sahuesos to send out their root network. Instead, the cacti appear to rely heavily on dew for moisture and on the rocky substrate to anchor them. Rather than send out hundreds of tentacles, their roots intercept dripping condensation channeled down the ribs of the branches to the base. Even on nearly solid rock, however, the roots penetrate the tiniest crack to provide a firm anchor against gulf storms.

Sahuesos' remarkable reproductive success may be related to the fact that they are sexually strange or perverse: they are trioecious. Some individuals have male flowers only, some females only, and some are hermaphroditic—that is, flowers are bisexual. No other columnar cactus is known to have this polymorph sexuality. Sahuesos flower in April and May. The large white flowers (up to 10 cm wide) open at night and are pollinated primarily by *Leptonycteris* bats that visit the flowers repeatedly throughout the night to drink the constantly replenished nectar.[53] The flowers close the following morning, never to reopen. In the early 1990s, I had the fortune one April to venture for a couple of nights to the sahueso-rich desert near Bahía Kino, Sonora, with bat experts Ted Fleming and Merlin Tuttle. We carried out a nocturnal vigil, peeping through night vision scopes as the bats flitted into view, slipped into a flower, slurped some nectar, then darted away into the darkness to visit other flowers, all in roughly one second. The flowers continue to produce sweet nectar throughout the night, seducing a bat into making multiple visits and assuring reproduction as its head parts the anthers and receives a dusting of pollen while the tiny creature laps at the sweet nectar. At the next flower, the bat flits in and sheds a smidgen of this dust on the stigma, and sooner or later each flower is pollinated. Thus do the bats ensure the perpetuation of their food source, and the sahuesos assure their own reproductive success. Fleming and Tuttle fitted some flowers with wires that prevented the bats from entering into the flower's nectar chamber and dusting the stigmas with pollen from other plants. The flowers fitted with these *excluders* for the most part remained unfertilized and failed to develop into fruits.

The earliest fruits ripen in late May and usually are in abundant supply by mid-June. The mature buds turn brownish and become covered with a mass of felt. From this thick, velvety cover, a few hundred relatively benign bristles emerge. They can prickle and irritate, but do not puncture the skin as do the fierce spines of most other cactus fruits. When the fruits are ripe, the skin or husk ruptures to reveal a bright purple interior. It is then a race among many competing organisms to gather the fruits. Birds—especially doves—and flying insects have the inside track, but hosts of ants also climb the giant trunks to the flowers and scoop out the luscious pulp. Once the fruits fall to the ground, they are consumed by another host of more terrestrial creatures, including coyotes, javelinas, foxes, and humans.

Although the sahueso's life span is impossible to determine with current methods, testimony from Seri Indians, who maintain long oral traditions and are familiar with individual plants, suggests that three hundred years is not unreasonable. Seris track the history of several ponderously large cacti near the village of El Desemboque, recalling that the giants were associated with deceased Seris well known in their oral history and with contemporary individuals as well. An ancient Seri custom is to bury an infant's placenta at the base of a columnar cactus, usually a sahueso. As the infant grows, he or she remains aware of the special significance of that particular cactus. Felger and Moser write: "The placenta of a newborn was buried at the base of a *cardón* [sahueso] or saguaro. Five small plants of any species were buried with it. Ashes were put on top of the place of burial to keep coyotes from locating it. The cactus served to mark the spot. In later years one might visit the site of his placenta burial to put green branches of any plant on it for good luck. Most of the people did not know when they were born, but each knew the general area where his placenta was buried."[54] Seris also identified notable plants with individual people.

Shreve estimated that sahuesos typically grow 10 cm in height a year.[55] Thus, an 18-m-tall plant would be around 180 years old. I suspect that Shreve's growth rates are generous and that the actual growth varies greatly with rainfall and local conditions.

A tiny population of mature and large sahuesos grows on hillsides of an interior mountain range in Sonora where rainfall exceeds 300 mm, or 12 in. In no other part of its range does average rainfall exceed 200 mm. Throughout much of the sahueso's range, rainfall is little more than 100 mm, even less in some areas. Usually, though, especially within a few kilometers of the coast, heavy dews drench the plants nightly for nine months of the year, an unmeasured source of moisture apparently critical to the plant's well-being. In cooler weather, dews near the Gulf of California begin to fall by late afternoon as temperatures drop rapidly. By sundown, all exposed surfaces are thoroughly wet, and the dew continues to collect throughout the night. As the dew condenses, it is channeled down the ribs to the base, where it can be absorbed by roots.

After years of contemplating the Sonoran population of sahuesos, I have concluded that the plants are probably recent arrivals to the area. This notion occurred to me in 1996, when Mexican ecologist Alberto Búrquez and I discovered a few large and mature sahuesos growing in rich tropical thornscrub in a canyon of the rugged Sierra Libre, nearly 60 km inland from the Sea of Cortés.[56] We quickly hypothesized that the Seris had brought them there. Whether the plants were there by accident or design, their size is compatible with an age that would place their birth or early youth roughly 250 years ago. This time frame would coincide with the Seris' occupation of the range for two decades in the mid-eighteenth century as a refuge when Spanish military forces besieged them. The sahueso is (or was, until recently) the Seris' most important plant, and bringing fruits with seed or even young plants would fit in well with their documented activities. (Fig. 2.70.)

It has also struck me that the sahuesos' range appears to be expanding, and on certain uninhabited and ungrazed islands in the Gulf of California densities of youthful plants are extraordinarily high. I consider it likely that Seris brought the plants to the Mexican mainland from Baja California, where their forebears may have originated. But, for the moment, pending DNA analysis and other population studies, this surmise is based merely on anecdotes, casual observation, and my familiarity with the Seris and their intriguing practices.

The Seris and sahuesos were at one time well nigh inseparable. Although numerous native peoples relish cactus fruits, sahueso fruits in Sonora until recently were the exclusive province of the Seri (and of those who learned from the Seris!). The historic range of the Seris has a remarkable overlap with the Sonoran distribution of *Pachycereus pringlei*, and the Seris' heavy reliance on the sahueso for food and implements is not surprising. Nabhan suggests that the Seri practice of placenta burial may be the source of the sahuesos in the Sierra Libre. He suggests that

Figure 2.70. Sahueso branches looking upward. They are probably the most massive branches of any columnar cactus.

Seris would transplant cacti and bury placentas under the transplants.[57]

Seris ate the sweet, delicate-flavored fruits in huge numbers. The ripe sahueso fruits bear fewer spines than most other columnar cacti fruits, so although the husks are thick, harvesting is less hazardous, and transportation is simpler. Collecting was carried out with a bacote, or long pole, made from a long agave stalk, lashed sahueso ribs, or any long, straight stick. A point was fashioned and lashed to the bacote, and the collector, usually a woman, wiggled the point around the base of the fruit (nearly always more than 3 m from the ground). Children would wait below and catch the falling fruits to prevent their becoming damaged or soiled because water for washing was at a premium in most of Seri and sahueso territory. At times, women would collect the sweet pulp and seeds in a container and discard the husks, but often the entire fruit was carried to the camp or village.

The time of ripening of fruits was also traditionally a time of rich harvests to which the Seris looked forward. In June 1970, I came upon a group of Seris on foot collecting fruits several miles north and inland of El Desemboque, one of their two villages. (At that time, Seris had no automobiles.) They had stuffed themselves with pulp and were carrying back to the village a couple of sacks bursting with ripe fruits. I gave them a ride in my Land Rover. They were a happy bunch, for the harvest that year had been bountiful, indeed. In spite of the roughness of the road, they laughed and joked—all eleven of them—clear back to the village. Their families greeted them with joy, for at that time there was no greater treat than the fresh pulp of the sahueso.

As was the case with other edible fruits in sunny climates, the Seris dried the pulp, saving the leathery mass, which would keep forever, they say. When the time came to eat it, it was chewed dry or soaked or boiled.

A small group of indigenous people related to contemporary Seris lived on the tiny island of San Esteban in the Gulf of California, some 12 km southwest of Tiburón Island. Bowen reports that although the sahuesos of the island produced abundant fruits, the San Esteban people did not especially like them, but still made wine from the juice.[58] I have not tried any of the fruits from San Esteban Island sahuesos, and it may be that the race on that island yields fruits of inferior flavor. Otherwise, I can only conclude that the San Esteban people were culinary philistines or that Bowen's informants were experiencing lapses in memory. Nevertheless, Hodgson cites a Jesuit missionary from Baja California who considered the fruits of sahuesos to be "nothing special" though he apparently had not tasted them.[59] Some observers have suggested that south of central Baja California pitaya dulce (Stenocereus thurberi) fruits were preferred to those of the sahueso, perhaps because the former were generally sweeter and easier to consume than the latter. I recall that my earliest impression (from the late 1960s) of sahueso fruits was that they were of indifferent flavor, inferior to those of pitaya agria (Stenocereus gummosus), pitaya dulce, and the saguaro. In retrospect, I can only conclude that my initial samples were insufficiently ripened or were from plants yielding fruits of inferior flavor. Maybe my youthful taste buds lacked maturity. Felger and Moser note that Seris identify four different colors of pulp and three different kinds of sahueso fruits, one of which produces bitter fruits.[60] Perhaps I tried some of the latter and formed an unpleasant opinion of sahueso fruits in general based on an inadequate sample. In the intervening years, I have consumed gobs of them and have found them uniformly excellent. At any rate, I can truthfully report that sahueso fruits collected in 1998 from the vast grove on a relict sand dune at El Cardonal, Sonora, were exceptionally sweet and tasty.

Each fruit contains up to two thousand seeds.[61] The average number is around thirteen hundred.[62] The seeds are soft, the size of very small peppercorns, and quite appetizing. The Seris formerly separated them from the pulp, carefully cleaned them, and preserved them in great numbers. Several reports state that Seris and Baja California native peoples formerly collected undigested sahueso seeds from dried human excrement, cleansed them fastidiously, and prepared them for repeat consumption, the so-called second harvest. This practice was understandable in the very dry and food-poor deserts of the region, for the seeds are an excellent source of fat and protein.

The Seris also stored huge numbers of seeds in knee-high pottery ollas to tide them through difficult times, sealing the wide mouths with lac resin. They deposited some of these ollas in caves and grottos as food caches. This Seri practice is the only instance I have managed to document in which the seeds of a cactus were afforded bulk long-term storage. Mayos still clean, sort, and save seeds from the etcho (Pachycereus pecten-aboriginum), a close relative of the sahueso, for domestic use, but the consultants with whom I have spoken have no recollection of storing them in containers as a cache. The seeds of the genus Pachycereus are larger than those of most other columnars, and the labor expended in concentrating the seeds made (and still makes) economic sense. The seeds' long-term storability and the Seris' practice of harvesting and storing them were perhaps the closest the Seris came to being agriculturalists. I suspect that they may also have deliberately planted seeds of some trees known to produce more or better fruits or both. If they did so, they were among the first horticulturalists in the New World. No accurate estimates for the Seris' historical tenancy of their lands have been attempted. Many of us who have spent considerable time with them believe that they have been in the area for several, possibly many, millennia. Perhaps they are direct descendants of the first peoples in the New World.

In the past half-century, many Mestizos (Mexicans of mixed Spanish and Native American background) have come to live in formerly Seri lands. They, too, harvest the fruits, enjoying the rich, sweet flavor and the crunchy nuts. It is ironic (and, I think, sad) that Seris no longer gather the fruits, preferring instead processed foods purchased at stores in their two villages—Punta Chueca and El Desemboque. In 2000, I took some Seris to the monte (bush) near El Desemboque, where we gathered a few buckets of the fruits. We ate as many as we were able—a couple of dozen each—and carried the buckets laden with fruit back to the village. According to the gatherers, it was the first time in several years that anyone had brought fruits to the village, in spite of the fact that they grow abundantly only a few hundred meters from the last Seri house.

The sahueso (called *xaasj* by the Seris) was by far the largest plant in most of Seri country (and in most of Baja California as well). In addition to the critical role of sahueso fruits as food, the cactus played an important cultural role in the Seris' lives. According to Griffen, Seris believed it could be made to rain if juice from the *biznaga* (*Ferocactus wislizenii*) was poured into a hole at the base of a sahueso.[63] The Seris found the ribs and trunk of dead sahuesos useful for making their shelters. Moran found the same use among Californios (descendents of the early European settlers in Baja California).[64] The wood of the trunks is surprisingly strong and hard, though not long-lasting. It is widely used by ranchers in Baja California to construct buildings—both walls and roofs. The ribs are strong enough when laid down sideways in a tight row to serve as a (more or less) comfortable bed. Other ranchers reportedly fashion furniture from the wood, but I have not seen the results of their carpentry. Finally, the great plants in Seri country are invaluable sources of shade in the pounding heat of the long summers. Near El Cardonal, a closely planted grove of sahuesos (my guess—I don't have any proof that they were planted!) forms a cavelike shelter that offers shade at all hours of the day. The trees' great size and age suggested to me as I lounged inside that I was only the latest of a long line of opportunists who appreciated those ancient wandering souls who had the foresight to plant these sahuesos in a tight circle. (Fig. 2.71.)

Seris would also wrap dead children in dense thickets of spiny plants and place this coffin on a secure branch of a sahueso to prevent coyotes from despoiling the burial site. One of the Seris' games or cultural practices involved forcing a half clamshell or other talisman into the flesh of a sahueso. The cactus would quickly form a tough callus around the foreign object, and there it would remain for the life of the cactus. Several sahuesos in the vicinity of Desemboque were thus permanently decorated, and the clamshells and assorted sticks cemented in the branches as though held in by a vise can be seen today. I have watched the sahuesos for thirty years. In the late 1960s, a sahueso near Desemboque was "decorated" with a rusting, worn baby stroller. I initially viewed the arrangement as a gag (the Seris have well-developed senses of humor), but later was told that the child that had been pushed around in the stroller (an American castoff) had been associated with that sahueso.

Figure 2.71. Ring of sahuesos, El Cardonal, roughly ten years after the photo in fig. 3.5.

Numerous biologists and laypeople have expressed concern about the fate of the saguaro and the native sahuesos of Baja California.[65] But relatively little such published concern has been expressed for the Sonoran sahueso. Where these cacti are able to escape persecution by humans, they appear to be doing quite well, even expanding their range in their highly restricted environment. Turner suggests that the absence of rodents and larger herbivores that prey upon seedlings dramatically increases survival of seedlings in island environments.[66] Turner, Bowers, and Burgess documented an astonishing increase in density of sahuesos on a small island near Guaymas: it "involved a population of a few old plants in 1903. By 1961, the population instead included many young plants. . . . [I]n 1964 the population consisted of 5,836 plants, or 8,000 per hectare. Many of these had become established between 1930 and 1960."[67]

Recent studies by Bashan and associates in Baja California Sur, however, have documented an increasingly common pathological condition among cardones there, a disease called "flat top" that causes branches to break off,

Figure 2.72. Sahueso and boojum trees (Fouquieria columnaris), *Cataviña, Baja California.*

leaving a scarred stump.[68] In some individuals, all or nearly all the branches are affected. The causes are unknown, but researchers have expressed concern that the illness may be spreading to other parts of Baja California Sur.

The origin of the name *sahueso* is obscure. Sobarzo, often cited as an authority on Sonoran place and plant names, traces it to the Cáhita (Mayo and Yaqui) *hahueso*.[69] However, the plant grows only marginally in the area historically occupied by Yaquis and is not in their current vocabulary. It is absent from Mayo country. Seris call the plant *xaasj*, not even close to the name *sahueso*. (Fig. 2.72.)

Pachycereus pecten-aboriginum (Engelmann) Britton & Rose

etcho (Cáhita), *cahue* (Tarahumara), *chiquí* (Guarijío), cardón, aborigine's comb

The etcho has the broadest distribution of any columnar cactus. It ranges from about 30°N 110°W near Opodepe, Sonora, on the Río San Miguel, southeast to 15.6°N 96°W on the southernmost coast of the Mexican state of Oaxaca, a straight-line distance of nearly 2300 km. Its range then extends east into Chiapas and may even reach into Guatemala. The plants fare well across an altitude gradient ranging from sea level to at least 1000 m. They grow to within a dozen kilometers or so of the Zapotecan archeological sites in the Valle Central de Oaxaca. Although they are seldom, if ever, found in dense groves, as are pitayas and tetechos, under favorable conditions a hectare in dense tropical deciduous forest may contain more than 250 individuals of varying sizes.[70] Etchos are common in coastal and foothills thornscrub at the northern end of the species' distribution. Howard Scott Gentry suggested this plant as a gauge of the plant community in the Río Mayo country of northwest Mexico: if the etchos grow taller than the surrounding trees, the vegetation is thornscrub; if the surrounding trees surpass the tips of the etcho, it is tropical deciduous forest.[71] To the south, the etcho (popularly called *cardón* outside of southern Sonora) is a common plant of dry tropical forests along the Pacific Coast and inland to where the dry forests merge into cloud forests. The great plants seem to have an affinity for tropical deciduous forest, for they are to be found throughout the west coast of Mexico wherever the dry forest is present. They are sufficiently abundant in southern Sonora that Mayos named one of their principal towns after them—Etchojoa, "Place of the Etchos." (Fig. 2.73.)

Etchos are common in southern Baja California Sur, but only in thicker vegetation in those few spots where rainfall exceeds more than a few inches each year. Where they grow, the very common *P. pringlei* tends not to be found. The two cousins overlap in many places, yet outside these mingling points one species prevails, and the other dwindles in numbers. Etchos compete well with trees and tall shrubs in the dense thornscrub of the Sierra de la Laguna, but sahuesos do poorly. Near the mostly abandoned village of El Triunfo, the state government has set aside a dense grove of etchos, which grow along with *Stenocereus gummosus, S. thurberi,* and *Pachycereus [Lophocereus] schotti.* Called Santuario de los Cactuses, this grove is one of the few places in Mexico where dense growth of columnar cacti is officially recognized and protected. No sahuesos are to be found in the grove, for it is located toward the bottom of an arroyo and is associated with several tall leguminous trees. On hillsides less than 1 km away, however, the desert scrub is more open, and sahuesos happily intermingle with the etchos.

Figure 2.73. Etchos (Pachycereus pecten-aboriginum) *in fruit, Masiaca, Sonora.*

Depending on rainfall, the etcho commonly reaches 9–10 m in height, occasionally more, in the northern parts of its range, although heights do vary. In coastal Oaxaca, in the extreme southern part of its range, it may reach a height of 15 m or even more. Measurement of some of the very tall specimens there is difficult due to the density of the surrounding vegetation. The trees have a habit similar to *P. grandis* but are of a trimmer aspect than the more massive *P. pringlei* and *P. weberi,* their branches growing more tightly pressed to the main trunk, especially in the south. Etchos frequently display a straight trunk that ascends 1 to 3 m (up to 5 m in the south) before branches develop. The branches usually protrude from the trunk at acute angles, forming a tight thicket of arms, all vertical. (Fig. 2.74.)

During the spring drought in tropical deciduous forest of northwest Mexico, etchos may be the only green plants visible on the parched hillsides. (Pitayas, *Stenocereus thurberi,* also common in the region, tend to turn a pale yellowish green.)

The etcho plants may tolerate light frosts, but their distribution suggests that they suffer or die where heavier frosts or freezes occur. At the very northern extremities of their range, they grow on southeast-facing hillsides where rising warm-air currents afford protection against frosts, and early morning rays of the sun heat the air quickly and melt any frost that has accumulated. Etchos (along with pitayas) farther south migrate to valley floors and become numerous on the coastal plain south of Guaymas. In the far south, they are seldom seen in significant numbers and, although common, usually form only a minor, if spectacular, component of the tropical deciduous forest in the region.

The buds are dark red and appear year round, but especially in late winter and early spring. Etchos flower almost randomly throughout the year, but their principal season in the north is January through March. In Sonora after a wet October 2000, millions of etchos produced myriad buds and flowers by late November. Mayos, natives to the region, assured me, however, that in spite of the early activity, the fruits would not develop until May and June, and they were correct. Some physiological reaction apparently prevents maturation of fruits until the amount of daylight reaches the right proportion. Some of the fruits ripen in May, but June is the primary time of harvesting.

A Mayo pointed out to me an etcho that had light golden buds. They indicated that the fruits would be whitish, he said. This rather rare form of the cactus has fruits with nearly white spines and white pulp (called *zarca* in Spanish or *tótosi* by Mayos), although the seeds remain black.

Etcho fruits are distinct—roundish and covered with hundreds of golden brown spines 2–5 cm in length in a mass that from a distance makes the fruits appear hairy. The dried floral tube remains attached until the fruit splits open at maturity, giving the fruit an unvarying appearance of the eye of the Cyclops. The fruits cluster around the upper portions of the upright arms, often so densely that they conceal the dark green cuticle beneath. Although the spines appear to be intimidating, they are rather weak, far less damaging to human skin than spines of *Stenocereus* or most other columnars. With practice, one can delicately hold a fruit while splitting it open and scooping out the contents, and the spines seldom penetrate the skin. Still,

Figure 2.74. Pachycereus pecten-aboriginum, *coast of Oaxaca.*

these quill-like spines often make access to the fruit difficult, one reason why the fruits are harvested less than they formerly were. The fruits of *P. grandis* are similar, but the spines are fewer and stronger, more capable of penetrating the skin. The husks of etcho fruits are much thicker than those of *Stenocereus* fruits and require considerable effort to split. Indeed, the pulp can be downright difficult to extract, for the husks can be resistant to opening. In Oaxaca, the fruits are not consumed because the effort required to extract the pulp is regarded as not worth the reward, and the pulp appears to be drier and less sweet than that of fruits of more northerly etchos. (Fig. 2.75.)

P. pecten-aboriginum is a most useful plant, recognized as such over its vast range. Lowland Guarijíos of the Sierra Madre of southeastern Sonora and southwestern Chihuahua, who eat the fruits and seeds and use the flesh for medicine, consider the etcho (which they call *chiquí*) to be their most important plant. The trees are prolific fruit producers—up to two hundred per plant. The wood is used

Figure 2.75. Etcho fruit.

in a variety of ways, and the flesh has numerous medicinal uses. The tree provides cooling shade when deciduous trees are leafless in late spring, a powerful virtue in a land where subsistence farmers must clear the land each spring. Mayos and Yaquis also harvest the fruits.

When the fruits ripen, the husks split, partially exposing the bright red pulp. The fruits are gathered with bacotes, but are usually twisted from the branches rather then pierced and pried, as is the case with many other fruits. They are often eaten directly upon harvesting by everting the husks inside out and popping the pulp into the mouth, a procedure that requires practice. Gatherers fill buckets with the fruit for domestic consumption and processing. The pulp is said to be easier to remove from the husk if the fruit has sat in a bucket for a few hours. Older Mayos used a special tool called a *to'oro ahuam* (*cucharra de cuerno*, horn spoon) to force open the husk and extract the pulp. One side of the spoonlike tool was serrated to permit cutting recalcitrant fruit membranes. (Fig. 2.76.)

The preparer (usually a woman) gathers the pulp of many fruits into a mass and boils it until the seeds drop to the bottom of the mixture. She then wraps the cooked mass in a cloth or pours it into a colander to express the juice, which she then boils for several hours and reduces to a thick miel (syrup). This miel is often mixed with the pulp and eaten with tortillas as a sweet. As recently as the 1970s Mayos fermented the miel into wine. In 1998, Guarijíos informed me that a fellow in one village still produced wine, but the practice was disappearing.

Guarijíos and Mayos separate the seedy pulp from the miel. They wash the pulp thoroughly and strain it through a coarse cloth to separate out the black mass of seeds, which they then dry on a *petate* (woven mat of palm) in the sun. The seedless pulp is often fed to swine. Once the seeds are sufficiently dry, they are stored in a *morral* (woven handbag of agave fiber). When the family needs the seeds, they grind and mix them with *nixtamal* (moistened corn flour) to make tortillas, or they grind the seeds alone to make what Guarijíos call *jípoca* (*atol de semilla de etcho*), gruel rather like cream of wheat. The atol is still popular and is a staple made in many households. Cooks frequently add a few tablespoons full of the ground seeds to a glass of water for

Figure 2.76. Collecting etcho fruits, Teachive, Sonora.

a refreshing and fortifying beverage called *pinole*. Mayos and Yaquis also carefully dry the seeds for such purposes. Several older Mayos have reported to me with no little pride that they grew up eating tortillas de semilla de etcho (etcho seed tortillas). These tortillas are still eaten in more remote villages where supplies of white corn from time to time run low. Older Cáhitas often lament the gradual disappearance of tortillas made from etcho seeds. Older Mayo women report that they used the oily seeds to season *comales* (griddles) before cooking on them. (The etcho seed reportedly contains 32 percent oil.)[72] Atol de etcho is often administered to persons with fever of influenza or general malaise, as an indigenous equivalent of chicken soup. Tarahumaras reportedly brew the flesh of young etcho branches into a hallucinogenic tea *(wichowaka)*. Guarijíos and Mayos use the flesh as a poultice for insect bites and eat the flesh raw or cooked to cure cancer.

I have been puzzled by the lack of interest in etcho fruits in Baja California and in southern Mexico. Several Californios with whom I spoke expressed surprise on hearing that the fruits of the etcho and sahueso, both of which are abundant in southern Baja California Sur, are edible. None of them had tasted the fruits, even though they lived in areas where the fruits are readily available.

The plants readily take to vegetative propagation. With timely irrigation, cuttings from the plants grow quickly. Living fences *(setos)* of etchos are common in the Río Fuerte valley and in southern Sonora much in the way that fences of malinche (órgano, *Pachycereus marginatus*) are popular in central and southern Mexico and fences of pasacana *(Trichocereus atacamensis)* are used in Argentina. Some etcho fences contain plants in excess of 8 m tall, considerably taller than the growth habit of *P. marginatus,* indicating a long, successful career as a fence, dating from a time when barbed wire had not been invented or was prohibitively expensive. These fencerows are now popular sources of etcho fruits and lumber. Those near habitations are usually well guarded by the residents, who view them as an orchard requiring no maintenance.

The lumber of the etcho trunks and lower branches is durable and surprisingly hard, though relatively light. It is ideal lumber for boards and small beams. I have seen it sawed by hand and planed into sturdy planks and then used to construct doors, one of which can be seen in figure 2.77. A Guarijío woman in the village of Sejaqui has such a door. Guarijío musicians fashion harps from thin strips of the wood. In the tropical forests of central and southern Sonora, it is the preferred wood for making bed frames. The owner of one such bed in Teachive claims it to be more than one hundred years old.

Figure 2.77. *Door made of etcho wood, Sejaqui, Sonora.*

In southern Jalisco, the lumber is sawed into strips that are fastened overlapping to make a solid roof. Unfortunately for the plant, the best lumber comes from living trees that are felled. The skin and cortex are stripped from the woody skeleton, usually with a machete, and the wood is allowed to dry for several days. The trunk and lower branches are then sawed into the appropriate lengths for the needed lumber. Properly seasoned, it does not warp and can be nailed in place. In Oaxaca, the wood is especially prized as a roofing material, with cane or tules interwoven with long strips of the wood to form a tight cover. Mayo blanket weavers incorporate the etcho wood into their looms, using lengths placed upright in the ground as posts to hold the warp and fashioning tamping sticks *(sasapayecas)* and spinning spindles from the wood. One such tamping stick I examined, owned by Sra. María Soledad Moroyoqui of Teachive, had seen more than thirty years of daily use. Constant contact with the taut wool warp had worn deep serrations into the edges and had polished the flat sides of the blade to a silky finish. (Fig. 2.78.)

Figure 2.78. Sasapayeca (tamping stick) of etcho wood in loom, Teachive, Sonora.

The stems are reportedly treated to produce a black dye used in tinting the hair.[73] I have not verified this use, but the flesh of the plants of the genus *Pachycereus* does turn black, and *P. marginatus* is used for the same purpose.

Etchos appear to reproduce well, although seedlings in heavily grazed flatlands appear to be less numerous now than they were during the early 1990s, probably due to trampling by cattle. In these areas, recruitment is compromised, and the perpetuation of an abundance of plants is questionable. On hillsides, especially steep or rocky slopes less accessible to cattle, recruitment appears to be quite satisfactory. Young plants are hardy and grow rapidly—as

much as 30 cm a year, according to native observers. When land is cleared for pasture, both etchos and pitayas are usually spared because they provide shade and are a source of fruits for cowboys. These relict individuals seldom survive well in the absence of protective companion plants, however. Evolved to depend on the company of myriad trees in tropical deciduous forests, they are subject to sunburn when isolated and usually die after a couple of years.

The etcho's species name *(pecten-aboriginum)* is derived from the aboriginal women's use of the spiny fruit husks as combs. Although I was skeptical that a cactus fruit could serve in such a capacity, Sr. Vicente Tajia of Teachive dem-

Figure 2.79. Chico (Pachycereus weberi) *near Chila de las Flores, Puebla. The largest chicos are found in this small region.*

onstrated the technique, scraping off part of the spines and slightly inverting the fruit to provide a place for gripping it. I tried arranging my hair with this aboriginal comb with modest success. I recommend that the reader stick with mass-produced store combs and brushes.

Pachycereus species of southern Mexico are as different in appearance as species of *Neobuxbaumia* or *Stenocereus* are similar. Taxonomists have argued for several decades before agreeing that these species belong in the same genus, and the placement there of some is still debated. It is not surprising, then, that natives hardly recognize them as related. All are well known to residents of the habitats in which they grow, but distributions of two of them are so sharply limited that residents of villages outside their narrow range may be completely unfamiliar with them.

Pachycereus weberi (J. M. Coult.) Backeb. *tenchanochtli* (Nahuatl, Puebla), cardón, chico, candelabra

At a maximum height of up to 20 m—around 66 ft.—the spectacular, spreading tenchanochtli (as *P. weberi* is called

in Nahuatl-speaking villages) of southern Mexico may be the tallest cactus. Other competitors for the tallest are the chicomejo *(P. grandis);* or perhaps the clavijas *(Neobuxbaumia mezcalaensis),* saguaros, sahuesos, tetechos, or etchos; or maybe the rare *Stenocereus chacalapensis* on the southern coast of Oaxaca—all of which are known to exceed 15 m in height. But with the possible exception of *P. pringlei* in northwestern Sonora, the tenchanochtli is the most massive cactus and surely the most charismatic. Individual plants of *Isolatocereus dumortieri* growing in the Valle de Acatlán in Oaxaca and Puebla rival the tenchanochtlis in size, but the latter becomes the clear winner when the two are viewed side by side (which can happen only using photographs). *P. weberi* is also the one of the easiest of columnars to identify, presenting unforgettable spreading candelabra of incomparable dimensions. Its color is an agreeable bluish green. The width of its canopy may nearly match its height, and its symmetry presents a most agreeable appearance. (Fig. 2.79.)

The sheer bulk of *P. weberi,* its many tons of plant matter, relative lack of menacing spines, and gobletlike arrangement of up to a hundred arms around the central trunk make the plants attractive to human entry. Although the branches appear from a distance to be more or less evenly distributed on the trunks, in most individuals they tend to form an outer ring, leaving the inside of the ring relatively free of protuberances. Children and small adults (and larger ones who *can*) frequently climb into this denlike central portion of some plants. At times, strategically located plants function as a rick where hay, straw, or *tasol* (corn husks) are stored to keep the fodder out of the reach of marauding livestock. It is simple for a man to stand inside the cupola and hurl pitchforks of silage or straw to the ground. Loading the material up into the crotch of the monster cacti is a different matter. (See fig. 1.9.)

Distribution of tenchanochtlis is spotty. They are most common in the southern Valles de Tehuacán and Cuicatlán ranging southward into Oaxaca and west into Guerrero, where they are less common and do not attain the great stature of plants found farther east and south. They may be common in an area and then not seen for several miles, or they may disappear altogether. Populations of the huge plants are found in impressive numbers on low foothills near Calipan and San Sebastián Zinacatepec, Puebla, south of the Valle de Tehuacán. On the hillsides near Teotitlán del Camino and south into Oaxaca, they are also common, the most notable plants on the natural landscape. Veritable forests of them sprawl over the hillsides of northern Oaxaca in the region of the Cañon de Tomellín. A side canyon near the small town of Quiotepec, Oaxaca, contains a forest of giants more than 15 m tall, along with much taller individual plants.

As common as the great cacti are in southern Puebla and western Oaxaca, peasants of the Valle de Zapotitlán, 30 km southwest of Tehuacán, although intimate with more than a dozen species of columnar cacti, are unfamiliar with tenchanochtlis. *P. weberi* appears to be quite sensitive to elevation, demanding conditions warmer than many other columnars in the region. In the higher slopes of the Cañon del Río Acatlán in Oaxaca and Puebla, it is absent, but massive *malayos (Isolatocereus dumortieri)* flourish, along with garambullos (*Myrtillocactus geometrizans*), jiotillas (*Escontria chiotilla*), chichipes (*Polaskia chichipe*) and several species of pitayos. Nearer the canyon bottom, however, and in the middle reaches of the canyon, malayos disappear, and huge tenchanochtlis become common indeed. Along the Acatlán, the latter are popularly called *chicos,* meaning "small ones." The closer one gets to the city of Acatlán, the larger the plants become.

On the pedimental bajadas east of Zinacatepec in the lower Valle de Tehuacán, plants of varying ages are flour-

ishing, and recruitment appears to be successful, with juvenile plants common. The chicos grow there in diverse thornscrub, a rich association with scrubby leguminous trees, jiotillos, garumbullos, pitayos de mayos (*Stenocereus pruinosus),* and a few xoconochtlis. They barely mingle, however, with the myriad tetechos growing on the adjacent steeper slopes. In contrast, near Calipan they grow on steeper slopes, but in isolation, and only mature plants are to be found. There appears to be little recruitment of young plants near inhabited portions of this area. The associated vegetation seems to have been deliberately cleared over the centuries and is now maintained in an open condition by large herds of goats, which are sure death to seedlings of all cacti and most other plants. (Figs. 2.80 and 2.81.)

The reddish chico fruits are thickly covered with spines, but are nonetheless edible, usually considered juicy and tasty by natives. The husks of the ripe fruits usually and conveniently split into four sections, rendering the juicy pulp rather accessible. Many generations of residents of the lower portions of the Valle de Tehuacán have apparently considered the fruits to be highly edible, for the remains of seeds have been found in coprolites (fossilized human excrement) from between 7000 and 8800 years ago. Because the cactus is often climbable, many fruits are accessible that in other cacti would remain tantalizingly out of reach even of an adroitly handled bacote. Still, such gathering sticks are assembled to wiggle the fruits from the heights of the great cactus. The seeds themselves are also eaten separately. The fruits are used in Oaxaca to prepare a fermented drink. (Fig. 2.82.)

Even more important in a region from which larger hardwood trees were long since virtually extirpated, the wood is excellent for construction. Natives living in zones where chicos are abundant point out that the wood has to be strong to hold up such a ponderous weight. Near the Cañon de Zopilote, Guerrero, local artisans remove the flesh and cuticle from the massive bases of the plants and allow the wood to dry. They saw off the arms to produce a level plane. On these various arms they place a circle or square of heavy beveled glass, producing thus a handsome table. These items of furniture are marketed in Acapulco, where they are quite popular among wealthy residents, especially Texans, or so say the vendors. Although the artisans producing the tables assert that they cut only dead trunks, I did a little sleuthing and found some massive carcasses of recently cut plants on which the green flesh remained. This is the somber side of an appealing craft industry. (Fig. 2.83.)

Few published accounts detail the use of tenchanochtli fruits for wine making, but they are widely used in Oaxaca

Figure 2.80.
Forest of chicos
(Pachycereus
weberi)
and babosos
(Pachycereus
hollianus),
Teotitlán del
Camino, Oaxaca.

Figure 2.81. Forest
of chicos, Cañon de
Tomellín, Oaxaca.

Figure 2.82. Fruit of Pachycereus weberi, *Cañon de Tomellín, Oaxaca.*

Figure 2.83. Table base of chico wood, Cañon de Zopilote, Guerrero.

to prepare a potent beverage called *vino de chico*.[74] During the months of May and June, one can often find *pulque de pitaya* in villages near Cuicatlán in La Cañada. It is also brewed from *P. weberi* fruits. I have not tried it, but when some becomes available, I shall.

The great trees are important sources of shade and protection from heavy rains. The larger ones, visible from afar, are individually known as though they were monumental and ancient trees. They sport trunks as thick as those of old oaks. They occasionally can be seen growing within fenced yards or corrals, where they shelter herds of livestock. New branches grow toward the periphery of the candelabrum, those nearest the center being the oldest. Old cuticle tends to scar over, either from age or from abuse, and turn dark gray.

This species was long included in the genus *Stenocereus*. However, studies with scanning electron microscopy revealed that its cells lack silica bodies characteristic of the latter genus and that the spines of its fruits resemble those of *Pachycereus pecten-aboriginum* more than those of any *Stenocereus*. Furthermore, the flesh, when cut, turns black at the margins of the incision. These and other factors led Gibson and others to refer it to the genus *Pachycereus*, where it is now comfortably at home.[75]

Tenchanochtlis are magnificent trees and should be revered among the world's giants.

Pachycereus hollianus (Weber) Buxbaum
baboso, acompés

Babosos are as thin and frail in appearance as chicos are stout and powerful. They often are as ugly as chicos are handsome. The plants—typically a single skinny stalk but often with a few tightly branched arms—reach a maximum of 6 m tall. The thin trunks often emerge from the ground in clusters of wildly varying heights. The thickets and individuals plants usually include senescent or dead trunks and arms. Some plants will appear half-dead, but they still grow. The result makes for a ratty-looking cactus, hardly an attractive plant, but strangely appealing. The cacti are allowed to grow in waste places and hillsides near human habitation. The ease of gathering the tasty fruits outweighs any inconvenience caused by the unattractive thickets, however. (Fig. 2.84; see also fig. 1.45.)

P. hollianus has a strikingly limited distribution. Early published accounts limited its occurrence to the northern portion of the Valle de Tehuacán and its tributaries in Puebla, but it grows in dense groves in the company of *P. weberi* near Calipan, Puebla, and in a roadside canyon south of Teotitlán del Camino, Oaxaca. (See fig. 2.80.) I rather think it grows in the canyon of the Río Acatlán and in the Cañon de Tomellín in Oaxaca, but have not found it in either place. Still, its distribution is confined to the greater Valle de Tehuacán, making it one of the narrowest endemics in the region. Its limited range is doubly odd in that it grows enthusiastically within the Valle de Tehuacán, reproducing prolifically and expanding promiscuously throughout that strange and wonderful canyon complex.

Figure 2.84. Dense grove of babosos (Pachycereus hollianus) *near Calipan, Puebla.*

The white flowers and purple, red, or yellow oval fruits emerge from the tips of the branches, resembling duck eggs stood on their ends. The tallest plants are said to produce the best fruit. The sweet, juicy fruits mature in July and are harvested and sold in markets. They are easier to eat than the fruits of the other *Pachycereus* species, with fewer and shorter spines that, though still numerous, are more easily removed. The pulp is more similar to the pulp of pitayo fruit, juicy with small seeds, than to that of other members of the genus.

In the Valle de Zapotitlán, babosos are widely planted to form living fences and corrals. The trunks grow densely and very close together, and the lower portions of the trunks often sport central spines up to 8 cm in length that typically emerge from the areoles in a generally downward direction, forming a ferociously intimidating barrier. A peasant from Zapotitlán showed me a corral completely fenced with babosos nearly 6 m high. He estimated the age of the corral at nearly one hundred years. The only access was through a gate that covered a gap in the fence. Such fences are ideal for goats, he said, and adequate for cows. The plants look rather fragile, but the central spines appear so intimidating that if I were a cow, I would not attempt to plow my way through. (Fig. 2.85.)

The wood of the lower trunk is solid, not latticed as is the case with most other cactus wood, with a hollow center. When the plant dies or is cut down, the trunk is sometimes dried and split and used for narrow roofing beams. It is ideal for this purpose because the trunks are uniformly straight, and the wood is quite sturdy.

The trunks of the baboso contain copious quantities of mucilage, which natives gather, apparently by making an angular slice in the trunk and collecting the dripping liquid. They boil the stuff to thicken the mucilage and market the product as an adhesive.

Figure 2.85. Corral of babosos, Zapotitlán de Salinas, Puebla. The plants have formidable, downward-facing spines that deter the most stubborn of cows and goats.

Babosos are so profoundly integrated into rural life of the Valle de Tehuacán that unequivocally wild specimens are seldom encountered. I cannot vouch for having seen a single wild plant in my roamings of the valley and its environs. However, near Teotitlán del Camino in the southern limits of the Valle de Tehuacán, a dense forest of babosos and chicos, with occasional garambullos and pitayos, forms one of the most impressive growths of columnar cacti I have seen. The babosos there could well be wild.

Bravo-Hollis has made the interesting observation that the seminal cactologists Britton and Rose chose *P. hollianus* as the *type*—that is, the defining species for the genus *Lameirocereus* (now *Stenocereus*).[76] It seems that the paradigm shifted, for baboso's characteristics clearly place it in *Pachycereus,* not in the genus *Lameirocereus,* for which it had been the paradigm. That genus needed a new type. Of such logical oddities are taxonomic decisions often made. *P. hollianus* will probably find its final home in a different genus.

Pachycereus marginatus (DC.) **Britton & Rose**
malinche, chilayo, órgano, jalacatito, Mexican fence cactus

The malinche, as it is widely known, is common throughout central Mexico from San Luis Potosí south to the Isthmus of Tehuantepec at a wide variety of elevations, from sea level to above 2400 m. It apparently grows in the wild: I have seen what I took to be wild plants in adjacent canyons of Hidalgo's Barranca de Metztitlán, in the Cañon Infiernillo, and in the Isthmus of Tehuantepec, all in isolated parts of the monte where cultivation would be improbable. Yet it is so commonly raised horticulturally that one can never be sure that a specimen found in the bush is wild. Indeed, it is grown in patios, yards, and fence lines in tens of thousands of households in central Mexico, especially in the highlands of Hidalgo and Querétaro, where it is probably the most popular ornamental plant. In the central highlands of Oaxaca, many miles of country and urban lanes are lined with the plants. (Fig. 2.86.)

The plant's typical color is a clean yellow green, with varying shades in different locations. It usually grows as a single trunk. When the malinche branches (probably as a result of injury), the new arms usually adhere in a tight and parallel fashion to the main trunk. The plant has comparatively few spines or none, the number varying widely over the distribution of the plant. I have seen specimens growing up to 7 m tall, but they are uncommon. An orchard in the Sierra Gorda of Querétaro has an unusual grove of the straight-stalked plants cultivated in an arrangement that resembles a pyramid. The species also does rather well horticulturally in the desert Southwest of the United States and in Southern California. (Fig. 2.87.)

Figure 2.86. Newly built fence of malinche cuttings (Pachycereus marginatus), *Mitla, Oaxaca.*

Figure 2.87. Horticultural aggregation of malinches, Sierra Gorda, Querétaro.

Figure 2.88. Fruits of malinches, Acatlán, Puebla.

P. marginatus grown as a living fence does triple duty also as an ornamental and an orchard plant. In the Valle Central de Oaxaca, new fences are constantly under construction, and nurseries raise and market plants for such purposes. Neat stacks of cuttings 1–2 m long can be seen here and there. Once the cuttings take root, they grow rapidly, reaching 4 m within a few years. Because the trunks grow rigid and very straight, they can be planted closely together, resulting in a tight, inexpensive barrier. Wire is sometimes woven back and forth between the cuttings to hold them in place until they can stand on their own. After that, the fence is permanent, pretty, and productive, for the trunks produce edible fruits along the ribs. In Puebla and Oaxaca, cuttings of pitayos de mayo and xoconochtlis are sometimes integrated into the malinche fences, providing fruit crops over a longer time period.

The red to pink malinche flowers are strictly diurnal and are pollinated primarily by hummingbirds.[77] The fruits ripen throughout the summer, producing reddish orange pulp (sometimes yellow as well). The buds are also eaten sautéed or chopped up and stewed. Due to the relative shortness of the trunks, the fruits and buds are easily harvested. In the Valle Central de Oaxaca, most of the plants are spineless, but do not bear fruits, whereas in neighboring Puebla the plants have spines and bear fruits, as do most of the plants to the north. (Fig. 2.88.)

In the northern part of the plant's range, the flesh of malinche trunks is dried and boiled to produce an intense black dye. In Mixquiahuila, Hidalgo, an Otomí city, women use a shampoo made from the boiled flesh of the cactus (which is called *órgano* there) to camouflage *canas* (gray hairs). A woman reported that after a tragedy in her family her hair began to turn gray. She cooked up some malinche shampoo and used it every day for a week. Her hair turned black from the pigments in the cactus and has remained so to this day. In the city, several pharmacies carry a bottled shampoo containing concentrate of the boiled cactus flesh. The shampoo is black and marketed under the name Orgánico. In the Valle de Tehuacán, residents swear that a similar preparation will cause hair to grow more swiftly and densely.

Figure 2.89. Grove of chicomejos (Pachycereus grandis), *Coyula, Oaxaca. These massive plants appear to be the only representatives of the species for nearly 200 km (124 mi.).*

In addition, when the hot weather arrives in the Valle de Tehuacán, cattle owners will lop off a malinche branch, chop it up, and stir the *machaca* (pulpy remainder) into water. It is widely believed that if livestock drink this potion, it will afford them protection against numerous ailments associated with hot weather. Others claim that if human consumers drink it, they will be rewarded with the same benefits.

The succulent flesh of malinches, as is the case with all members of the genus *Pachycereus,* turns black when cut. Gibson and Nobel have used this feature as diagnostic of the genus.[78]

Pachycereus grandis Rose
cardón, chicomejo

The distribution of chicomejo, perhaps the least common member of the genus, appears to be limited to a small area south and west of Mexico City in the warm, semiarid canyons of the states of Mexico, Morelos, Oaxaca, and perhaps Michoacán. In these areas, it grows mostly on steep canyon sides in tropical deciduous forest (or what was forest until it was cut down) and is frequently quite inaccessible. The plants grow very tall, perhaps the tallest of all cacti, with reports of individuals in excess of 25 m in height—more than 82 ft.!—hence its specific name *grandis*. I hope that such tall individuals exist, but I rather doubt that these cacti grow much higher then 15 m. The plants resemble the etcho, with very long branches growing in close proximity to the axis of the trunk, so confusion with the latter is possible, as I can testify. The fruits are quite different, however, the husks more felty and less spiny, and the pulp drier and less palatable. Peasants in the Río Acatlán valley assured me that only livestock eat the fruits. The wood, however, is very strong and excellent for roof beams, they said.

A remarkable grove of very large chicomejos grows on a small plateau near the mountain Oaxacan hamlet San Pedro Coyula, in a valley tributary to the Valle de Cuicatlán. They are robust plants, very large, and recognized in the village for their beauty and, to a lesser extent, for their fruits, from which pulque de pitaya, a cactus wine, is brewed by villagers from the town of Quiotepec in the valley far below. I looked forward to having a taste, but in the previous week four Quiotepecans had been murdered (probably by other villagers), and the probability of my also being murdered was too high for my liking. (Figs. 2.89 and 2.90.)

In the lower Acatlán valley, chicomejos can be found

Figure 2.90. Pachycereus grandis, *Coyula, Oaxaca. The fruits are tasty, but the size of the plant makes harvesting them a challenge.*

growing alongside etchos, chicos, and malinches, making the region a veritable cornucopia of *Pachycerei.* The valley, which empties into the Balsas basin, is thoroughly populated, but in some places retains good examples of original vegetation. The best growths of *P. grandis,* however, are to be found on the convoluted cliffs of southwestern Morelos. The Coyula population grows in the drainage of the Papaloápan basin that empties into the Atlantic.

Pachycereus [Lophocereus] schottii (Engelmann) Britton & Rose

sinita, senita, *musue* (Cáhita); *hasahcápöj* (Seri); old man cactus

Sinitas (often spelled *senita* in English) are plants of the warmer desert valleys of the Sonoran Desert, spreading into adjacent thornscrub to the south, formerly well into Sinaloa, where they may have once been common. (The habitat in northwest Sinaloa in which they would have grown has been mostly bulldozed and flattened for irrigated agriculture.) A few individuals can still be found on the bajadas and volcanic hillsides near Topolobampo on the Sinaloan coast. They are abundant in arid Sonora and in Baja California, flourishing in dry arroyos in the Pinacate Volcanic Range of northwestern Sonora, where rainfall is less than 75 mm annually. The plants appear to thrive under extremely hot and very dry conditions. Sinitas in the Pinacate Volcanic Range appear to be robust even though the area may see more than a year without measurable precipitation. They seem to do equally well in thornscrub of southern Sonora, with a mean annual rainfall in excess of 300 mm. Sinitas appear to dislike each other's presence, for apart from thickets (a result of the sinita's propensity for vegetative reproduction), the plants tend to be widely dispersed. Although large thickets of the plants are common, sinitas do not seem to form forests. (Fig. 2.91.)

These Sonoran Desert stalwarts possess two peculiar traits. They produce multiple flowers from long-bristled areoles (the areoles of *Myrtillocactus* also bear multiple flowers but are free of such bristles), and they send down a taproot to complement the network of horizontal surface roots. The closest relative of the sinita appears to be *Pachycereus marginatus,* the malinche, of central Mexico, so based on this relationship (and on other details) researchers have referred it to the genus *Pachycereus.*[79] I am conservative about changing names. I would rather leave this old friend and most engaging cactus in its more historical genus, *Lophocereus.* However, its similarity to the malinche forces me concede it to *Pachycereus.*

P. schottii relies heavily for pollination on *Upiga virescens,* a pyralid moth whose larvae live on its fruits and seeds.[80] This is the first known case of mutualism in columnar cacti—that is, symbiosis between a plant and an insect. The plant cannot pollinate well without the moth, and the moth cannot survive without the plant. The cactus sacrifices about 25 percent of its fruits to sustain the moths, but it gains a faithful pollinator. It is a small price to pay. (Figs. 2.92 and 2.93.)

The sinita is the least common of the three columnar cacti found in the United States, growing only near the Mexican border in Organ Pipe Cactus National Monument and in a couple of nearby desert ranges, with only a few individuals to be found in these places. Immediately south of the border, sinitas become numerous and well known, suggesting that the plant much prefers living in Mexico than in the United States. Over much of its range,

Figure 2.91. Healthy sinita (Pachycereus [Lophocereus] schottii), Sirebampo, Sonora.

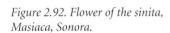

Figure 2.92. Flower of the sinita, Masiaca, Sonora.

Figure 2.93. Grove of sinitas, El Cardonal, Sonora.

its arms, though vertical, are short, usually less than 3 m tall. Occasionally, however, they grow much taller. A subspecies, var. *australis*, from southern Baja California reaches 8 m in height, making it a tall columnar. I have seen individuals more than 5 m tall growing in thornscrub on the Sonoran mainland and in more moist portions of the desert, but plants of this height are rare on the mainland. In the Cape Region of Baja California, the sinita forms most impressive associations with etchos and sahuesos.

A bizarre genetic variant, var. *monstruosus*, whose branches are covered with large bumps and creases, is found in only one canyon in the north-central portion of Baja California. Dubbed the totem pole cactus, it readily propagates vegetatively and is quite popular in the United States as an ornamental. A specimen growing in my yard in Tucson, Arizona, grew in fifteen years from a 30-cm-high single column to a branched monstrosity more than 4 m tall. Yet another strange form, var. *mieckleyanus*, is found in a separate canyon not far away. A southernly variant, *P. schottii* var. *schottii* forma *spiralis*, with a corkscrew rib pattern, grows in Baja California Sur. It is diminutive, appearing to reach little more than 2 m in height.[81]

P. schottii is easily identified in the field, for the areolas of the upper portions of the branches are densely covered with long bristly gray spines (often a rich rufous color in Baja California), so much so that the entire upper part appears gray bearded, hence the English common name "old man cactus." Plants commonly grow in thickets exceeding 10 m in diameter, consisting of many vertical arms emerging from ground level. In southern Sonora and Baja California, plants with trunks are not unusual.

Cuttings of sinita propagate readily. Indeed, given enough rainfall, fallen arms left in place nearly always send out roots, and before long branches sprout, and another clone begins its individual life. A lopped-off branch randomly cast to the ground will often send down roots and develop into a large cactus. Natives of the region in which sinitas grow casually take a machete to a branch and plant the cutting in their yards as an ornamental. This wild desire to send down roots leads to the growth of the dense thickets. Seedlings are less common, but are readily identifiable in the field, beginning as a button. It is difficult to determine which mode of propagation is more important to the species.

The white to rosy white flowers (rarely yellow)[82] open at night beginning in spring and continuing well into midsummer. Fruits appear in July through late October. The fruits, somewhat smaller than a golf ball, are usually without spines and are tasty. The plants produce relatively few of them, however, so fruits are not gathered, as are those of pitayas and sahuesos. Furthermore, the extensive thickets with many branches can deter the hopeful fruit gatherer, and where sinitas grow, pitayas are usually also present. The pitaya fruits are larger and tastier, so in Mayo country, at least, sinita fruits should be considered a snack food, but are often left to birds, coyotes, foxes, and insects. Among the Seris, however, with fewer wild food resources than Mayos, the sinita fruits were more important. Felger and Moser observed that Seris would partially lop branches with a knife or machete to make them droop.[83] The manipulation would then make them grow in a more downward direction and render their fruits more easily accessible to harvesting.

Among Sonorans, the flesh and skin of the sinita are frequently boiled into a tea that is drunk to relieve the pain of arthritis. A tea brewed from just the flesh or pith is sometimes promoted as a wash for cleansing the body. In Sonora, it is widely held that medically effective material must come from a branch with five ribs, not six or seven. The five-ribbed variety, which is uncommon but abundant in certain localities, is also believed to have cancer- and diabetes-curing properties. Some plants have both five- and six-ribbed arms.

Sinitas were especially important to the Seris in addition to their use as food. Felger and Moser observed the Seris' use of columnar cacti over a period of more than thirty years. They note that Seris believed that sinitas, which the Seris called *hasahcápöj*, could be used to incorporate spiritual advantage over others—namely, to put a hex on an enemy. They also acknowledged, however, that the use of the plant for manipulation of others' luck could backfire on the user. At a puberty festival I attended in the early 1980s, women and girls sat in a circle in the sand playing a game in which cross sections of the cactus roughly 10 cm wide were used as tokens. I also noted that dried sections of the same width with the outer flesh removed were joined through the center with an ocotillo or saguaro rib and used as toys by little boys, who would scoot them along with a stick. One fellow even fashioned an ersatz truck from the wood of *joopingl* (*Bursera hindsiana*) using two of the sinita toys as axles and wheels. It was by no means a streamlined vehicle, but Seri lads have great imagination and the liveliest inventiveness at games; they enjoy them more than any other people I have known.

Pachycereus [Lophocereus] gatesii (M. E. Jones) D. R. Hunt

Although I have searched for this admirable cactus, I have not seen it in the wild. It grows only on the west coast of

Figure 2.94a. Tepamo (Pachycereus tepamo), *Palo Pintado, Michoacán. The blue-green color sets this species off from all others in the region.*

Figure 2.94b. Tepamo near Infiernillo, Michoacán.

Baja California Sur in the vicinity of the town of Todos Santos. It does rather well horticulturally, however. In the gardens of the late H. B. Wallace of Scottsdale, Arizona, a fine plant that reaches more than 3 m in height produced a cutting that grows in my yard. Its many spines give it a characteristic appearance. I have neither a common name nor any ethnobotanical information about it.

Pachycereus tepamo sp. nov.
tepamo

During my first visit to Mexico's lower Balsas valley in 2002, I was intrigued by what I took to be a somewhat diminutive form of the chico (*Pachycereus weberi*). Local residents referred to it as *tepamo* and noted that the fruits were quite good to eat and the branches made decent lumber for ceilings. In my notes, I recorded it as *P. weberi*, writing that it appeared to lack the massive habit, displayed less resemblance to a candelabrum, bore fewer ribs, and had more prominent areoles than the plants of Puebla and Oaxaca.

It was not until early 2004 that I discovered that a Mexican and a Chilean had already described the tepamo as a new species.[84] Indeed, it was so different that I wondered how I could have confused it with the chico. The authors of the monograph thoughtfully named the species after its local name, maintaining an excellent tradition of perpetuating indigenous names for plants in the scientific nomenclature.

The tepamo is a most handsome plant, the young growth nearly blue in color, with crisp lines and a goblet shape. I did not see individual plants more than 8 m in height and noticed them only within a very limited area around the village of Palo Pintado, Michoacán, a hamlet that seems to be a focal point of cactus evolution. In 2006, I found huge plants grown in the Cañon Infiernillo. I suspect that as tepamos are more adequately studied, they will be found over an even greater area. Until the time when numbers of the plant can be demonstrated sufficient to ensure survival, *P. tepamo* should be considered a threatened or endangered species. (Figs. 2.94a and 2.94b.)

Tetechos and Their Allies: The Genus *Neobuxbaumia*

The fascinating genus *Neobuxbaumia* has not been adequately studied, probably due to its members' relative isolation from large cities or urban areas. Nine species have been described, and at least one, but probably more, remains to be described. They represent a variety of habits and reproductive patterns, all (except perhaps for *N. euphorbioides*) displaying a characteristic appearance, including many ribs, light green color, and often a faint cap or topknot formed of spines, lending to most species its own color of headdress. All are found in southern Mexico west of the Isthmus of Tehuantepec, with only *N. euphorbioides*, *N. polylopha*, and possibly *N. squamulosa* extending north of the latitude of Mexico City. And nearly all species, except for *N. euphorbioides*, produce individuals that grow to become very tall cacti. The genus has a proclivity for producing species with extremely narrow distributions. In all probability, more species will be discovered with the expansion of botanical knowledge of southern Mexico's rugged southern coastal ranges and the wild slopes of northern Oaxaca.

Neobuxbaumia tetetzo (Coulter) Backeberg
tetecho

The tetecho is the most spectacular species of a spectacular genus. *Neobuxbaumias* in general and tetechos in particular are among the tallest of cacti. Though the tetecho grows only on highly restricted sites, it is the most numerous of the eighteen species of columnar cacti found in Mexico's Valles de Tehuacán and Zapotitlán of Puebla and in the closely adjacent Valle de Cuicatlán (Río Grande Quiotepec) in Oaxaca. Tetechos abound also in great numbers in the Cañon de Tomellín, Oaxaca, and along the Pan American Highway connecting Oaxaca and Tehuantepec, where they probably reach their greatest height.

Although these populations are the best known and documented, scattered smaller populations also appear along the Oaxacan coast. On limestone cliffs near the Chontal village of Santiago Astata midway between Huatulco and Salina Cruz, large numbers of tetechos jut far above the rolling bajada through which the highway runs. I first noticed them during the rainy season, where they were visible only on the cliffs. I had not read about this population, but their zen dictated that they were a *Neobuxbaumia*. When I returned later during the dry season and finally located accessible individuals (previously obscured by the riot of green, jungly summer growth), they turned out for all appearances to be tetechos. An additional population, also rather sparse, grows near San Pedro Juchatengo along

Highway 135 connecting Puerto Escondido and the city of Oaxaca. (Figs. 2.95, 2.96, and 2.97.)

Along the foothills above the valley floor near the Pueblan towns of Calipan, Zapotitlán de Salinas, and San Sebastián Zinacatepec, tetechos often grow in groves (tetecheras) of nearly impenetrable density. So numerous are the cacti in these forest pockets (as many as eighteen hundred mature plants per hectare) that the foothills appear to be painted with their light green tint. The Pueblan tetecheras, all within a 70 km radius with Tehuacán at the center, end and begin abruptly. Immediately adjacent to nearly impenetrable forests are similar habitats without a single tetecho. The species is apparently confined to the valleys and canyons of Oaxaca and Puebla, suggesting that its ecological requirements (apparently including some sort of calcareous soil) are narrow. (Figs. 2.98 and 2.99.)

At Zapotitlán de Salinas, Puebla, the Mexican government has designated a stand as part of a biosphere reserve and has established a botanical garden nearby. Not only do tetechos grow in truly unbelievable numbers, but they are also among the tallest of columnar cacti. Individuals grow in excess of 15 m tall, as can be seen in figure 1.6. In the Cañon de Totolapan, southeast of Oaxaca, they grow even taller—perhaps as much as 20 m. The trunks and arms grow typically straight, sometimes with arms branching off near the apex. Their branches (usually few) tend to grow parallel—often tightly parallel—to the usually straight trunk. In the tetecheras, there is often no room for extra branches, and many individuals are merely single straight trunks.

From a distance, the tetechos' resemblance to saguaros is striking. Indeed, their similarity is more than superficial. Several authorities place the two giants close together phylogenetically. Although tetechos are often somewhat skinnier than saguaros, in a mixed grove, saguaros would be difficult to distinguish from either *N. squamulosa* or the tetecho except during flowering and fruiting seasons. I suspect that some day *Neobuxbaumia* and *Carnegiea* will be merged into one genus. Wallace, however, finds genetic reasons to maintain the two genera.[85]

In some habitats, straight-trunked tetechos are also easily confused with clavijas (*Neobuxbaumia mezcalaensis*). A spectacularly thick forest of the two species intermingled swathes a hillside near San Juan Raya, southwest of Zapotitlán de Salinas. Clavijas tend to grow in a single trunk, but many tetechos do as well. About the only way to differentiate the two is to look for the tetechos' bristly white cap, created by the numerous long white spines at the apices of the trunk and arms. In clavijas, the white cap is absent. When the two plants are in fruit, however, it is simple to

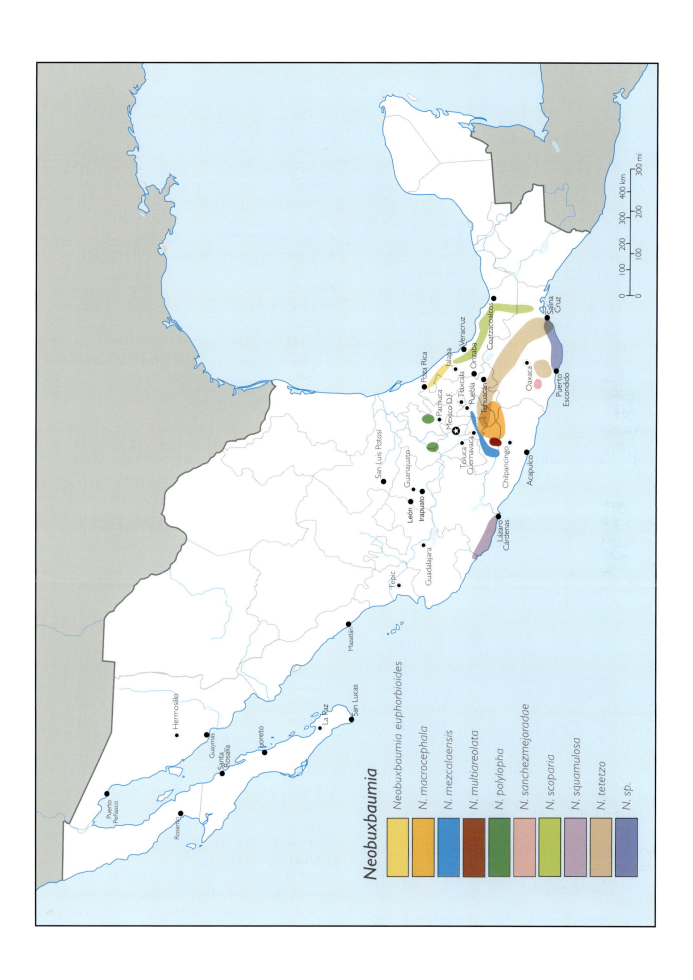

Neobuxbaumia

Neobuxbaumia euphorbioides

N. macrocephala

N. mezcalaensis

N. multiareolata

N. polylopha

N. sanchezmejoradae

N. scoparia

N. squamulosa

N. tetetzo

N. sp.

Figure 2.95. Tetechera (forest of tetechos, Neobuxbaumia tetetzo) *near Zapotitlán de Salinas, Puebla.*

Figure 2.96. Tetechera, Cañon de Tomellín, Oaxaca.

Figure 2.97. Tetechera near Calipan, Puebla.

Figure 2.98. Tall tetecho near La Reforma, Oaxaca. East of the Valle Central de Oaxaca, the tetechos seem taller and thinner than those of the Valles de Tehuacán and Tomellín.

Figure 2.99. Inside a tetechera, Zapotitlán de Salinas.

distinguish them, for the white flowers and the fruits of *N. mezcalaensis* proliferate wildly along the ribs below the crown of the cactus for nearly half its length, whereas in tetechos they are confined to the crown and just below it, as is the case with the saguaro. The tall clavijas are more widely distributed than are tetechos, being especially prevalent in southern and eastern Morelos and along the Río Mezcala in Guerrero, where they grow scattered in great numbers on hillsides.[86]

Opinions vary among villagers in the region as to the rate of growth of these great cacti, but all agree that tetechos grow slowly. Some natives maintain that they grow so slowly that they are not useful for living fences, as are babosos (*Pachycereus hollianus*), malinches (*P. marginatus*), and xoconochtlis (*Stenocereus stellatus*). Peasants have estimated to me that the oldest tetechos are more than two hundred years of age, a longevity that would roughly correspond with that of the better-documented saguaro. These natives' testimony should be given weight, for they have resided in the same villages in the region for generations and have thus been able to track older cacti from one generation to another. Whether tetecho growth is substantially increased by irrigation has not to my knowledge been tested. In the closely related saguaros, growth is increased dramatically with supplementary watering.

If we look at the locations where tetechos grow well, we can begin to understand their narrow ecological requirements. The annual rainfall at Zapotitlán de Salinas, Puebla, at 1500 m elevation is roughly 400 mm, nearly all of which falls between May and September.[87]

Data from the eastern Oaxaca populations are limited, but the similarity of vegetation in the two environments suggests that annual rainfall is similar, perhaps slightly higher in Oaxaca. The vegetation of the slopes frequented by tetechos in Puebla is heavily overgrazed thornscrub, with few trees matching the height of the vast numbers of columnar cacti. At Zinacatepec, Puebla, 1150 m elevation, where tetechos also appear in great tetecheras, annual rainfall is 535.5 mm, but the area is somewhat lower and hotter and subject to greater evaporation than in the region of Zapotitlán de Salinas. The vegetation in the lower valley is transitional thornscrub and tropical deciduous forest, with well-developed tropical forest in protected canyons. Totolapan, Oaxaca, on the Cristóbal Colón Highway, lies significantly lower at 850 m and receives 460 mm annual rainfall. All three areas are frost free, and the tetechos flourish on steep hillsides down to more gently sloped pediments (bajadas). The coastal population grows in dense tropical deciduous forest where rainfall exceeds one meter annually, but only on limestone outcropping. The inland population along Highway 135 also grows on

limestone, but emerges from scrubby vegetation. The densest stands of tetechos are confined to the Valles de Tehuacán, Zapotitlán, and Cuicatlán. (Fig. 2.100.)

In heavily grazed areas, recruitment of tetechos is poor, and few young individuals can be found, the fate of many large cacti in Mexico. In other areas, however, especially on lands under the control of indigenous communities, grazing has been limited in the densest forests, which the local people view as places of great importance, even sacred. The indigenous Popolocas of Zapotitlán initially viewed the government-created botanical gardens as an invasion of a sacred place and profess to have been perplexed and even angered by the government's unilateral action in creating the reserve. They claim that the groves were well protected under their community guardianship. Members of the community now staff the gardens and provide guide service, including explanations of the tetechos' virtues.

On lower foothill slopes where overgrazing has occurred, the tetechos grow with greater space between individuals, and recruitment has been slower. Over the years, however, viciously thorned catclaw (*Mimosa luisana*) has grown rapidly, filling in the open spaces with a vengeance, discouraging the passage of all but the lowliest of mammals. This growth may afford protection to tetecho seedlings and juvenile plants. In the denser groves, grazing would not be practical at any rate, for without substantial felling of the cacti, even goats would find grazing between closely placed, great-spined giants a hazardous undertaking.

Why the tetechos grow in such striking groves is difficult to explain. Paleoecologist Tom Van Devender suggests that over the centuries the relentless persecution of rodents as a food source by native peoples of the region may have eliminated or at least greatly suppressed an important source of cactus predation, allowing huge numbers of plants to survive the seedling stage when they are most vulnerable to gnawing predators who relish the tender shoots and young cacti.[88] The same hypothesis may explain the extraordinary proliferation of sahuesos on tiny islands in the Gulf of California. The Valle de Tehuacán has been inhabited for more than seven thousand years, and humans have been exploiting local plants and animals for that time, so determining how the vegetation occurred "naturally" is virtually impossible. No other cacti in the region are found in such densities, so we may assume that tetechos love society.

The tetecho flowers in February through May, and the fruits begin to mature in May, corresponding with the onset of the rainy season in southern Mexico. Two species of bats appear to be its only pollinators. A variety of organisms (including humans) assist in seed dispersal.[89] (Figs. 2.101 and 2.102.)

Figure 2.100. Inside a tetechera,
Calipan, Puebla.

Figure 2.101. Tetecho flower,
Zapotitlán de Salinas, Puebla.

Figure 2.102. Crested tetecho, San
Pedro Juchatengo, Oaxaca.

The subspecies *nuda,* diminutive and nearly free of spines, can be found growing not far west of Tehuantepec, Oaxaca, on the limestone slopes of the Sierra Guiengola at an elevation of less than 100 m and on limestones along Highway 135 near San Pedro Juchatengo. The Tehuantepec plants are far shorter than most tetechos, but retain the typical tetecho habit. The scarcity of spines causes the areoles to stand out, presenting a mosaiclike appearance for the cactus. I have not seen the flowers or fruits. (Fig. 2.103.)

Residents of the Zapotitlán area view tetechos as a great resource. In the past, many natives of the region gathered cactus products in profusion during the late spring. Although the practice is declining, residents are still fond of the buds, flowers, fruits, and seeds. Buds scrambled with eggs remains a popular dish, and cooks of Zapotitlán still add flowers to soup broth. Most homes still stock jars of buds pickled in vinegar; they are tasty. The highly edible fruits—called *saleas* (Puebla), *zalehitas* (Oaxaca), or *higos de teteche* (tetecho figs)—are harvested in great numbers, eaten fresh, and preserved. Their flavor is quite different from most other cactus fruits. Only mildly sweet (similar in taste and texture to saguaro fruits, but less sweet), they are reminiscent of partly dry figs and possess a distinct nutty flavor. I found the fruits to be tasty indeed, perhaps more as a food than as a sweet. (Fig. 2.104.)

Tetecho fruit production differs from many other species in that it varies dramatically from year to year. Natives of the region report that if rains come early—in January or February—few flowers will develop, and the fruits that mature will then rot on the branches. If the onset of rains delays and allows the fruits to ripen, the harvest can be excellent. Bumper crops occur only once every five years or so. Saleas (the ripe higos de teteche) split open, exposing the brownish fruits, which can be gathered by probing the tops of the branches with long poles (bacotes). Some saleas are dried and saved. In addition to the fruits, the seeds are separated from the pulp and eaten in their own right.[90] Large batches are combined, and the fruits are washed to leave the seeds behind. The seeds, rich in oil, may be eaten uncooked or roasted and ground into rich flour. Popolocan cooks often add this ground seed meal to fried chilis to make a fiery local salsa. The seeds are also a key ingredient in local *pipián,* a chile and pumpkin seed–based salsa.

According to Popolacas, the wood until recently appeared in most households in the form of vigas (crossbeams) and firewood. Prior to the planting of eucalyptus and the availability of commercial lumber in the area, the straight trunks, called *cuilotes* (Oaxaca), fulfilled most applications requiring lumber and were the only straight trunks available. Some tetecho roofs remain in the Valley de Zapotitlán,

Figure 2.103. Neobuxbaumia tetetzo *var.* nuda *on limestone in the Sierra Guiengola, near Tehuantepec, Oaxaca.*

Figure 2.104. Salea, fruit of the tetecho. Only mildly sweet, the fruits have the flavor and texture of dried figs.

but in spite of their durability, such roofs are considered the mark of low classes and a home with a *techo de tetecho* is considered a mere hut. Although residents of Zapotitlán assured me that houses with tetecho ceilings and roofs still exist, no one could or would point one out to me.

In spite of the huge numbers of tetechos in southern Puebla and northern Oaxaca, the genus *Neobuxbaumia* and its (at least) ten species, has not received extensive international attention, perhaps because the genus appears to be confined to very limited habitats in central and southern Mexico far from the haunts of most cactologists. Three additional species of the genus *Neobuxbaumia* are found in the same region as the tetechos. Yet another, *N. multiareolata* (Dawson) Bravo, known only from hard-to-reach limestone outcroppings in the adjacent state of Guerrero, may also be found in southwest Puebla. It has purple flowers and a dark green trunk. A recently described species, *N. squamulosa* Scheinvar et Sánchez-Mejorada, is found near the Pacific coast of south-central Mexico in the states of Colima, Guerrero, and Michoacán.[91] *N. sanchezmejoradae* occurs only on limestone outcroppings in a restricted portion of a canyon in the Mixteca Alta. *N. polylopha* of Hidalgo, Querétaro, and Guanajuato is more accessible, but is restricted to limestone tallis slopes, which occur only in spotty locations and in quite limited numbers.

Neobuxbaumia euphorbioides (Haw.) Buxb.
pitayo

Although this smaller Neobuxbaumia species has edible fruits, it is the least recognized locally (and least distinguished) of the genus. Superficially it resembles a straight stalk of the far more common *Stenocereus griseus* and is not, to my knowledge, recognized locally as a distinct species. This is understandable, for I had a good bit of difficulty locating it and at first took it for another species. It is found in cut-over semideciduous tropical forest in southern Tamaulipas, extreme eastern San Luis Potosí and possibly northeastern Hidalgo and Puebla, as well as in the moist canyons of northern Veracruz below about 1000 m elevation. It achieves a height of about 7 m, but is usually shorter. It grows as a single stalk (except where it has been injured) and can easily be overlooked in the lush forests in which it seems happy. Bravo-Hollis reported it to be common in the Río Actopan valley,[92] but I found it only in a small area in the upper part of the valley west of Trapiche del Rosario and encountered no one who had anything to say about it. Within a few kilometers in the valley below grow large specimens of *N. scoparia,* and nearby are *Pilosocereus leucocephalus* and *Stenocereus griseus,* all of

Figure 2.105. Neobuxbaumia euphorbioides, *Río Actopan, Veracruz.*

which have a charisma that easily outmatches this rather insignificant cactus. (Fig. 2.105.)

Although the forests below Jalapa, Veracruz, are rapidly being destroyed, I was delighted to discover numerous seedlings of *N. euphorbioides* growing on basaltic rock (it appears to join *N. scoparia* in growing on basalts or other volcanics) in an area where the larger individuals had been cut down. The roughness (and high temperatures) of this surface probably discourage goats from frequenting the *malpais,* as the lavas are called, and the ever-regenerating canopy of this very moist area provides the necessary shade to guarantee recruitment.

This cactus is not spectacular. When flowering, however, it becomes noticeable, for the prolific pinkish flowers are larger and wider than those of most other *Neobuxbaumias.* The plants are dwarfed by other species in the genus, but deserve to be studied further under field conditions, and ethnobotanical investigations need to be continued to determine more of its humanly useful products, if any. In its habitat, *N. euphorbioides* must compete with a great diversity of tropical plants, making study of it an excellent opportunity to observe its place in tropical vegetation.

Neobuxbaumia macrocephala (Weber) Dawson
órgano de cabeza roja

This very large columnar is apparently endemic to the Valle de Tehuacán complex and is found there only in small numbers and on a few restricted hillside sites. A few are found on steep hillsides north of Zapotitlán de Salinas. Others are seen near Highway 385 southwest of Coapan; another group flourishes on hillsides near Reyes Metzontla, Puebla, and yet another, the largest concentration, on a pediment near Acatepec. Near Coapan, residents refer to the giant as *órgano de zopilote* (vulture organ). (Fig. 2.106a.)

The ecological requirements of *N. macrocephala* must be narrow, indeed, for its range to be so restricted—even more so than that of the stately *Mitrocereus fulviceps* with which they grow. Macrocephs are handsome cacti, perhaps more so than the more slender tetecho. They are also more massive, in excess of 12 m tall, with a few large, thick arms, some of which exceed the main axis of the plant in height. Flowers and fruits develop at the tip of the branches, similar to the style of *N. tetetzo*. The branches are usually capped with a burnished mop formed by dense, bristly rufous spines, hence its common name *órgano de cabeza roja*, "red-headed organ pipe." It is red only in comparison with the nearby *órgano de cabeza amarillo (M. fulviceps)*, whose cap is more golden, but to whom this worthy plant bears a superficial resemblance. The colors of the two may grade into each other, and when the two species comingle, one of them, usually *N. macrocephala*, may be overlooked. On my first visit to their habitat, I did not realize that I had encountered a "new" population of *N. macrocephala* on the high northern slopes of the Valle de Zapotitlán. Only when I scrutinized some of my photos several months later did it become apparent that two species were present. What I had taken exclusively for *Mitrocereus* in the field was a fine and healthy mix of two species. Rib count is a dead giveaway, however, with macrocephs sporting twenty-three or more ribs and mitros fewer than fifteen. The general orientation of *N. macrocephala* is also the more purely vertical of the two, and its branches tend to parallel and hug the main axis more closely. Mitros tend to be a lighter green in color than the macrocephs. Macrocephs appear to grow exclusively on calcareous soils, maybe on limestone only. (Fig. 2.106b; see also fig. 3.2.)

Residents around Reyes Metzontla consider the small fruits of macrocephs to be dry and tasteless, hence worthless, especially given the nearby proliferation of many

Figure 2.106a. Órgano de cabeza roja (Neobuxbaumia macrocephala) *near Coapan, Puebla. Note the reddish tint to the branch tips.*

Figure 2.106b. Neobuxbaumia macrocephala *near Acatepec, Puebla.*

Figure 2.107. Grove of Neobuxbaumia macrocephala *near Acatepec, Puebla.*

species of cacti with tasty or sweet and succulent fruits. However, they also say that plants growing on the northern slopes of the valley near Zapotitlán produce fruits sweeter than higos de tetecho and almost worth the considerable effort needed to harvest them from the tips of the branches. The lumber is said to be strong and useful, better than that of tetechos, but the scarcity of the plants has rendered lumbering impracticable, and the plants and cactologists who admire the macrocephs must be grateful thus to see them spared. These giants tend to be solitary or grow in groves of modest numbers, in sharp contrast with tetechos and clavijas.

In its habitat, *N. macrocephala* is also commonly found in a most agreeable association with several other columnar cacti, the barrel cactus–like *silla de suegra (Echinocactus platyacanthus)* and the lilies *sotolín (Beaucarnea gracilis)* and *izote (Yucca periculosa)*, all in an open setting. Although the landscape is forever under attack by livestock, especially goats, the scenery and the vegetation are quite simply splendid. The Zapotitlán Desert presents as fine an example of xeric vegetation as is to be found anywhere. Macrocephs

stand a good chance of surviving to expand their range over the next few eons. (Fig. 2.107.)

Neobuxbaumia mezcalaensis (Bravo) Backeberg
clavija, clavellín, gigante, telephone pole cactus

The remarkable clavija ("pin") or gigante ("giant") is far more widespread than the previously described *Neobuxbaumias,* ranging well beyond Oaxaca and Puebla into the states of Colima, Guerrero, Jalisco (reportedly), Michoacán, and Morelos. It appears to prefer mountain slopes in tropical deciduous forest, growing in large numbers (but never in dense associations) on more moist hillsides interspersed with numerous short trees, especially numerous species of *Bursera.* Outside of its association with *N. tetetzo* and other columnars in the Valle de Zapotitlán (parts of which may have once been covered with robust tropical deciduous forest), the clavija is common in dry tropical forests where other columnars are absent, reduced in numbers, or confined to cliffs or rock outcroppings. The single-stalked giants reach a whopping 20 m tall (they *may* be the tallest cacti of all). They thrust through the canopy to tower high over the treetops like so many green telephone poles. (Fig. 2.108; see also fig. 1.7.)

Individual clavijas may develop arms, but apparently only in response to injury. In the lusher tropical deciduous forest along the highway between Acatlán and Izúcar de Matamoros, Puebla, clavijas are quite abundant and handsome, their light green contrasting agreeably with the darker green of the *Bursera*-rich tropical forest during the summer monsoon, like enormous pins jutting from a green pillow. The same landscape is duplicated in the Balsas depression. Along the Río Mezcala (for which the species is named), clavijas form an important part of the lush summer vegetation, but remain green in the dry season. They appear to prefer limestone to other substrata, but do not seem limited to it. (Fig. 2.109.)

Residents of the Valle de Acatlán have assured me that the flowers are edible and are cooked into soups. The fruits, black on the outside with white pulp, are sweet and widely consumed as well. They are also easier to harvest than those of other members of the genus *Neobuxbaumia* because they grow well down from the branch tips and are thus accessible. Natives report that the lumber is quite strong and indeed suitable for roofs. The long, straight shafts are ideal for roofing beams. It is fortunate for the plants that they yield edible flowers and fruits, for natives are probably reluctant to exterminate any plant that provides them with food so easily gathered. When the fruits ripen, they burst open like stars, the husks opening in three or four sections. Thus, from a distance they resemble white flowers. The buds turn black. (Fig. 2.110.)

Bravo-Hollis carried out a study of differing characteristics in separate populations of clavijas. She discovered that plants in these disjunct regions have differing characteristics, but the variation is insufficient to create a new taxon.[93] For example, the plants of the upper Valle de Acatlán flower in March and April and ripen in May and are thus locally called *pitaya de mayo,* while other populations may not ripen until June. The variations are clear evidence, however, of new strains that are evolving in the isolation of different canyons. In a few thousand years, we should have new species or two.

I am responsible for the English name of the plant and fully accept any criticism.

Neobuxbaumia multiareolata (E. Y. Dawson) Bravo Scheinvar & Sánchez-Mejorada
vela

This isolated species is confined to limestone outcroppings and cliffs in a small area in the state of Guerrero. The single-stalked plants, locally called *velas* (candles), flourish

Figure 2.108. A Neobuxbaumia mezcalaensis *20 m tall (66 ft.) near Huajuapan de León, Oaxaca. Note the similarity to the saguaro.*

in humid semideciduous cloud forest at nearly 1000 m on north-facing hillsides. The southern slopes are covered with pine-oak forest. At first glance, one would hardly guess that they are a different species from *N. mezcalaensis,* which are common in Cañon de Zopilote only a few kilometers to the north. However, in 1932 the Mexican cactologist Helia Bravo-Hollis noticed that these cacti bore reddish purple bell-shaped flowers, whereas *N. mezcalaensis* had funnel-shaped and primarily white flowers, rarely grading into whitish purple. Further study led to the naming of the velas as a distinct species.[94] (Fig. 2.111.)

The plants grow in excess of 10 m tall, straight as a telephone pole. I saw not a single plant with branches. I did not collect either flowers or fruits, for the plants show a decided proclivity to grow on inaccessible limestone cliffs, and the

Figure 2.109. Clavijas (Neobuxbaumia mezcalaensis) *in tropical deciduous forest near Izúcar de Matamoros, Puebla.*

Figure 2.110. Neobuxbaumia mezcalaensis *in fruit, Puebla.*

Figure 2.111. Vela (Neobuxbaumia multiareolata) *on limestone, Tierra Colorada, Guerrero.*

Figure 2.112. Neobuxbaumia multiareolata *rising from tropical deciduous forest in limestone substrate, Guerrero.*

probability of my plunging to my death in the collection effort was too great for my liking. I was able to photograph a plant with buds growing on a huge (and unclimbable by me) limestone boulder just south of Acahuitzotla on the old, winding Mexico City–Acapulco highway. Near Tierra Colorada, Guerrero, the plants can be seen (at a distance) on limestone. On granites, the dominant parent rock in much of the area, none is to be found, even on the north-facing slopes. (Fig. 2.112.)

The flowers resemble midget push-up underarm deodorant cylinders, purple-red on the outside with white reproductive organs on the inside located well below the tip formed by the petals. Local residents report that the fruits are not eaten. I suspect that this is more due to their inaccessibility than to unpalatability, for the fruits of other *Neobuxbaumias,* though dry, are often tasty and nutritious. Residents report that the *varas* (ribs) are often used as cross-hatching in roofing. I was unable to locate any roofs in which the varas had been used, but homes in the immediate vicinity of the plants are few and far between.

Neobuxbaumia polylopha (DC.) Backeb.
órgano, totem pole cactus

This órgano occurs only in a matrix of sharp-surfaced boulders on steep limestone slopes in arid canyons. It shares with *N. multiareolata, N. sanchezmejoradae,* and *N. squamulosa* a refusal to grow on any material other than limestone. Although I have read reports of populations in Hidalgo, Querétaro, San Luis Potosí, and Guanajuato, its occurrence is everywhere spotty. The isolated populations also tend to be poorly known, even to local residents. I have been unable to determine either a common name (other than the ubiquitous *órgano*) or any local uses. I have sampled the fruits and find them to have a delicate, nutty flavor, which should be shared with others. (Figs. 2.113 and 2.114; see also fig. 1.52.)

Although *N. polylopha* is poorly known locally, it is popular among cactus fanciers. In the huge Barranca de Metztitlán of Hidalgo, three populations of the plants rise from limestone tallis. The area has appropriately been

Figure 2.113. Totem pole cacti (Neobuxbaumia polylopha), *on limestone tallis, Barranca de Metztitlán, Hidalgo.*

Figure 2.114. Totem pole cactus looking upwards, Barranca de Metztitlán, Hidalgo.

designated as a protected area by the Mexican government, especially important in that it receives numerous visits from cactologists and plant enthusiasts from Mexico and around the world. For decades, many visitors came not just to view the cacti, but also to poach them. Rangers reported to me that foreigners have gone to great lengths to smuggle seeds and small plants out, even hiding seeds in their underclothing and shoes. (How their thieving ways were found out is not clear.)

Enthusiasm for this great cactus (and the remarkable *Cephalocereus senilis*, which also frequents the canyon) is understandable. The plants grow straight and tall—in excess of 15 m in height. They often have an attractive craggy appearance, sometimes produced by an abundance of ballmosses (*Tillandsia* spp.), sometimes by distinct knobs and external growth rings that give the plants individual appearances and suggest totem poles. A population on limestone tallis not far north of Metztitlán has roughly one hundred plants growing in an area perhaps three acres in size. The plants grow so tall that it is difficult to photograph them in their entirety. Recruitment of juveniles is excellent,

rebounding well since grazing by livestock (primarily goats) was curtailed by the government in the late 1990s. Local residents spoke of the possibility of nursery production of the cacti within the Barranca de Metztitlán, an activity that would quickly reduce the incentive for theft of seeds and small plants. No concrete plans for the enterprise had yet been established in 2004. (Figs. 2.115 and 2.116.)

This species reproduces well in nurseries. In southern Arizona, the plants grow quickly and tall, thriving in the heat and tolerating cool temperatures well, seemingly prospering in the region's calcareous soils. They are an ideal plant for landscaping in hot, dry climates where freezing is uncommon.

Neobuxbaumia sanchezmejoradae A. B. Lau *tunshichi* (Mixtec)

Evolution and typography have isolated this poorly known *Neobuxbaumia* even more than the narrowly distributed *N. multiareolata*. It appears to be the most localized within a genus of several localized taxa, centered some 100 km west of where *N. multiareolata* is found. It was first described

Figure 2.115. Fruits of Neobuxbaumia polylopha.
*In appearance and taste they are remarkably similar
to tetecho fruits.*

Figure 2.116. The tallest Neobuxbaumia polylopha,
approximately 16 m tall, Barranca de Metztitlán, Hidalgo.

by Lau from a 1976 collection.[95] It is known only from a population growing at around 1500 m near several pueblos in the traditional and isolated Mixtec *municipio* (county) of Santiago Nuyoo, Oaxaca, where it appears on rugged limestone cliffs and outcroppings (not, as reported by Lau, on eroded volcanic rock) near the base of a gigantic (1300-m-high) escarpment of apparently pure limestone. Lau believed the species has strong affinities with *N. mezcalaensis* and *N. multiareolata,* both of which are found within 100 km of the *tunshichi* population. *M. multiareolata* also grows *only* on limestone outcroppings (as does *N. polylopha* in Hidalgo and *N. squamulosa* along the coast and inland). *Mezcalaensis* tolerates nonlimestone substrata. (Fig. 2.117.)

Tunshichi plants grow as tall as 10 m, according to Mixtecs from Santiago Nuyoo (12 m according to Lau). The branches are somewhat more spindly than those of tetechos, but closer to 15 cm thick, half a foot, as opposed to the 10 cm observed by Lau. The specimens I was able to study (from a precarious perch) were about 8 m tall. The plant prefers the most inaccessible crags on the steep slopes and tends to cohabit with ample numbers of *mala mujer* (stinging nettle, *Cnidoscolus* sp.), which makes access even more difficult. I viewed the tunshichis at the end of the dry season in May, and the approach was still difficult. In the rainy season (the region receives more than 2 m of rainfall

annually, according to Lau), it would be impossible to get close to them due to the junglelike conditions.

Tunshichi is found on both sides (east and west) of the immense canyon that forms the upper reaches of the Río Atoyoquillo, an important tributary of the great Río Verde that empties into the Pacific some 150 km to the south of Santiago Nuyoo. Access to larger populations of the plant that were reported to me is very difficult, for they are found on the far side of the canyon, which is crossable only by a torturous footpath and requires a jaunt of several hours during the dry season. Access to Santiago Nuyoo requires a more than two-hour drive (dry season) on a dirt road of dubious maintenance that climbs as high as 2850 m in chilly pine forests before dropping 1200 m in the space of 10 km to the pleasant tropical valley. In the rainy season, the road is frequently impassable.

The Mixtecos with whom I spoke claim that tunshichi does not bear fruit. Lau, however, reported that in 1992 botanists at the Universidad Autónoma de México received a ripe fruit from a local student in Santiago Nuyoo and were thus able to provide the plant with a taxonomic description.[96] The flowers, apparently white or light pink, are located at the apex, as is the case with *N. tetetzo.* The fruits are surely edible, but commentary on them was not available in Santiago Nuyoo. The fruits of several other *Neobuxbaumias* are diminutive and might hardly be notice-

Figure 2.117. Tunshichi (Neobuxbaumia sanchezmejoradae) *near Santiago Nuyoo, Oaxaca. The plants grow on steep limestone cliffs and appear to be limited to a small number growing around this isolated village.*

Figure 2.118. Tamping stick made of tunshichi wood, Santiago Nuyoo, Mixteca Baja, Oaxaca.

Figure 2.119. Neobuxbaumia scoparia *on volcanic substrate, Río Actopan, Veracruz.*

able from a distance, and all the tunshichi plants grow at a distance from human habitation. The tunshichi has an identifying characteristic within the genus *Neobuxbaumia:* individual plants sport a thick tuft of black spines at the apex of the branches, giving the appearance of a topknot to the plants.

Although knowledge of the plant's reproduction appear to be poorly understood locally, other aspects of its life cycle are well known. The dead trunks are the source of the very hard, light wood used in fashioning tamping sticks for back looms used by Mixtec women of the region to weave the textiles they wear and sell in regional markets. The one pictured in figure 2.118 is from the hamlet of Union y Progresso, where similar looms have been in continuous use for decades and, apart from tiny serrations caused by being slammed against the weft untold thousands of times, are no worse for wear. The wood is clearly durable and is prized, no doubt, because of its hardness and its scarcity. (Fig. 2.118.)

From the descriptions by residents of Santiago Nuyoo, I doubt if there are more than a few hundred plants of this species, which represents an evolutionary mystery.

Neobuxbaumia scoparia (Poselger) Backeberg
órgano, pitayo

The common name of this fine cactus indicates the poverty of local nomenclature. *N. scoparia* is a large plant, indeed, often reaching more than 10 m in height with dozens of branches. Some Veracruz individuals reach 12 m tall and merit the label *giant.* (Fig. 2.119.)

These many-branched columnars grow in decidedly isolated and spotty localities from the Isthmus of Tehuantepec in Oaxaca northeastward to the Río Actopan valley northeast of Jalapa, Veracruz. They appear to flourish only on substrates of volcanic origin, which may explain their spotty appearance in the region. Rainfall in Tehuantepec, where the eastern plants grow, is decidedly seasonal, the tropical deciduous forest there receiving roughly 650 mm annually, while those plants near Jalapa grow in a habitat that grades from semideciduous tropical forest with rainfall in excess of 1500 mm to tropical deciduous forest with rainfall probably less than 1000 mm. The órganos appear to prefer the drier portions of this wildly variable rainfall regime. The Tehuantepec population is not to be found on the valley floor (perhaps because of complete deforestation there) but can be seen on the lowest slopes as soon as the volcanic rock emerges from the valley fill. Near La Ventosa on the eastern side of the isthmus, several plants skirt the base of a large limestone inselberg. Above them on rocky outcrops grow numerous *Cephalocereus nizandensis* plants. As is the case in the Río Actopan valley, some of the Tehuantepec plants grow in conjunction with *Pilosocereus leucocephalus* in a thorny forest that produces bloody scratches and punctures on those who are not on to its tricks. In Tehuantepec, I was gratified to locate the plants more or less precisely where Bravo-Hollis had indicated.[97] The population growing farther east has not been mentioned in the literature.

These órganos tend to develop a thickened growth of reddish spines near the apex of the branches, lending the upper parts a bushy, reddish cast. The fuzziness contributed by the numerous spines causes the plants to appear to be out of focus or perhaps painted over with the sfumato glaze of the Italian Renaissance masters. Although the specimens in Veracruz grow within a few kilometers of *N. euphorbioides* plants, I did not find them growing sympatrically and doubt if they do, for *N. scoparia* seems to prefer a drier regime than does *N. euphorbioides*.

I have found few people who acknowledge having eaten the fruits of *N. scoparia*. A Oaxacan fellow owned that they were edible, the pulp orange and not very sweet. Mostly, he said, the birds get them, and people eat the more accessible and tasty pitayas from the local *Stenocereus griseus* or *S. laevigatus* strain. It seems that the fruits are snack foods for cowboys and others who range through the hills. Pitayas are more prolific fruit producers, and their fruits are juicier and easier to harvest.

The *scoparia* wood, however, is said to be dense and useful in construction of houses. A cowboy in the vicinity of the Río Actopan praised the wood of the trunk and lower branches. He reported that it was popular in bygone days for house construction, especially useful for roofing timbers. It has a local reputation as a strong wood, and if the poles and beams were kept dry, they would last indefinitely and seemed immune to termites and insect borers. However, the advent of corrugated metal roofing and easily obtained pine beams has virtually eliminated the use of the lumber. The cowboy could not think of any homes in the region that actually had used the wood of *N. scoparia*. This decrease in human use is undoubtedly beneficial to the cactus, for Mexico's huge human population growth has affected this idyllic valley as well and would certainly endanger the viability of the local populations of this fine cactus.

Figure 2.120. Neobuxbaumia squamulosa *near Colima, Colima. Note the strong resemblance to the saguaro.*

Neobuxbaumia squamulosa Scheinvar & Sanchez-Mej.
nochtli (Nahua)

This rather large and tall cactus (reportedly up to 10 m in height), with none to few or many arms, grows on the coast of the states of Colima, Guerrero, and Michoacán in southwestern Mexico. It is reported from Sinaloa, but I question the accuracy of those reports. Along the Pacific Coast, it is found on southwest-facing limestone rock outcropping and cliff faces where populations of up to one hundred individuals stand out like matchsticks viewed from afar. Its range is substantial, but the habitat requirements are sufficiently narrow that the plant is more or less rare. It is apparently used extensively inland in Michoacán, so much so that the rather small populations there are endangered. (Fig. 2.120.)

I have searched for it unsuccessfully in Guerrero, where Bravo-Hollis reported finding it near Zihuatanejo.[98] Hillsides where she reported it have been cleared or are now covered with heavy second-growth forest that would probably crowd out any columnar cacti. To the west along the coast of Michoacán it appears dramatically on bare rock faces jutting from dense tropical deciduous forest, especially between Cachán and San José de Alima. Most of these populations are inaccessible except by means of miserable battles through the jungly forest. The area is hot during the dry months, and progress through the forest is slow, as one fights vines, spiny and thorny branches, and dense, prickly underbrush. In the rainy season, only madmen would think of trying to penetrate that green hell, glorious from a distance in its emerald boisterousness, but utterly hostile to human entry. I approached one cliff-dwelling population (in wintertime) only to have my way cut off by a dense grove of 2-m-high stinging nettles (*Cnidoscolus* sp., mala mujer), whose dry stems still retained thousands of tiny, threatening, venom-laden spines. A machete might have made it possible to reach the plants, but the limestone outcroppings from which they grow are rough with legions of serrated edges, and the cliffs are so steep as to deter all but the most foolhardy.

Occasionally the plants can be seen on gentler limestone slopes. Here they proliferate wildly, a heartening sign. Their growth appears to be stunted, however, probably due to the difficulty in obtaining abundant nutrition and root mass from the niggardly limestone stratum on which they grow. Few of these plants sprout arms. (Figs. 2.121, 2.122, and 2.123.)

Inland and at elevations between 200 and 500 m above sea level, especially in Colima on the highway between Tecomán and Colima City, the plants grow abundantly on limestone hillsides lush with tropical deciduous forest. In the dry season, they can be seen from afar, bearing an uncanny resemblance to saguaros, to which they are clearly closely related. Indeed, the flowers are white and cluster near the apex of the branches, as do those of saguaros, but, unlike the saguaro, *N. squamulosa* does not appear to bud from the crown. Its flowers have pinkish sepals, whereas saguaro flowers are white or green. Saguaros have rather more ribs—up to thirty, though usually fewer; *squamulosas* have seventeen or fewer. Still, the apparent similarity of the plants is remarkable, indeed. From a distance inland, *N. squamulosa* would be virtually indistinguishable from a saguaro if the two grew side by side. Only in two or three other cases do two different species of columnar cacti so resemble each other.

These inland plants may reach 8 m in height, perhaps more, and branch more frequently than the ones that grow along the coast. As is the case with the coastal populations, they seem delighted to spawn their young, and the forest floor is well furnished with juveniles of all sizes. In contrast to many of Mexico's great columnars, this species is reproducing extremely well. It is heartening to see large numbers of young cacti, from tiny seedlings to plants several years old, growing in the immediate vicinity of mature giants, although their distribution is spotty and limited. A plant nursery not far from Colima sells juvenile plants purportedly raised from seed.

This plant, called *nochtli* (the Aztec term for cactus) is well known to the Nahua-speaking peoples of the Michoacán coast, who find the flowers edible and much to their liking. The wood is excellent for house construction. Were it not for the species' high rate of reproduction, the widespread use of the wood might place the species in jeopardy. The plants' inaccessibility, their fondness for otherwise inhospitable limestone cliffs, and their success in producing progeny guarantee that the species will continue to flourish.

Neobuxbaumia sp.

In the year 2000, while I was searching for *Stenocereus chacalapensis* on the central Pacific coast of Oaxaca, I spied a large columnar cactus growing on volcanic rock among viciously thorny shrubs and small trees on a steep hillside near Huatulco. From its light green color, I immediately took it to be of the genus *Neobuxbaumia*, but could not recall any reports of the presence of that genus along the Oaxacan coast.

I managed to scramble up to the plant and found there were several, definitely *Neobuxbaumias*, including some juveniles. I later found more of them growing farther east in tropical deciduous forest along the coast. The tallest plants were about 7 m tall, resembling in form *Neobuxbaumia tetetzo*, but with clearly different areolar, spine, and rib patterns. They also closely resemble *N. squamulosa*, but have fewer ribs (thirteen as opposed to fifteen or more) and are of a somewhat darker green color. The fact that they were growing in volcanic rock and not on limestone seemed also to rule out *N. squamulosa*. I sent a photograph to a Mexican ecologist, who contacted taxonomists at the Universidad de Autónoma de México. He reported that the Mexican scientists have also noted the plant and have pronounced it to be a new species that they are in the process of describing. I await their findings. (See fig. 1.29.)

The plant grows in a very wild area, and I was unable to find anyone locally who was familiar with it or its uses. The juveniles bear a close resemblance to those of *N. euphorbioides*, though they are a much lighter green, and adults have naturally occurring arms, which *N. euphorbioides* lacks.

Figure 2.121. Juvenile Neobuxbaumia squamulosa plants, near Maruata, coast of Michoacán. The substrate is unforgiving limestone, but recruitment is obviously going well.

Figure 2.123. Flowers of Neobuxbaumia squamulosa. *Photo by Marco A. Morentín.*

Figure 2.122. Neobuxbaumia squamulosa *plants on rugged limestone. The plants appear to benefit from their inaccessible location.*

Figure 2.124.
Undescribed species of
Neobuxbaumia *near*
Huatulco, Oaxaca.

Landscapers at the resort region of Bahías de Huatulco have planted many of these handsome plants along road medians. The region sports abundant granite outcroppings, and local people report that the *Neobuxbaumias* grow on the granite successfully. (Figs. 2.124 and 2.125.)

Neobuxbaumia sp.

Lau reports having seen many *Neobuxbaumia* plants growing on a (limestone?) cliff in very moist cloud forests near San Juan Teutila in the Sierra Juárez of northeastern Oaxaca.[99] No further information is available. Given the extremely narrow requirements of other species of the genus, it is quite possible that Lau saw the only population of this species. I searched for it, but residents of the region assured me that the extremely remote Mestizo town of San Juan Teutila was situated in oak and pine, not in cloud forest, and there were no limestone cliffs nearby. I gave up the search at that point.

Figure 2.125. *Close-up of undescribed* Neobuxbaumia *near Huatulco, Oaxaca.*

The Hairy and Bearded Cacti: *Cephalocereus* and *Pilosocereus*

This group of hairy columnars includes the genera *Cephalocereus* and *Pilosocereus*. Although two genera are involved, they are closely related phylogenetically, and some taxonomists (at least for a while) have considered them one genus.[100] All the species discussed here have prominent white to cream-colored to gray woolly tufts growing on their upper branches or spread up and down the branches. In most cases, they exhibit what is called a *pseudocephalium*, a dense growth of wool or hair from which flowers and fruits emerge. A true *cephalium* (derived from the Greek *kephalos*, meaning "head") is located at the apex, and flowers and fruits emerge at that point. In this group, the fruits may be submerged in wool that resembles spun fiberglass insulation, but they are spineless. These plants grow in diverse habitats from northern Mexico well into Peru. I consider in this section *only* those species growing in North America and the Caribbean.

Cephalocereus

The genus *Cephalocereus* has five species, all Mexican. Three of the five have a narrow distribution on limestone slopes and cliffs in eastern Oaxaca and western Chiapas. The taxonomy of the group is still suspect, and we should expect changes as more phylogenetic research is carried out. In general, *Cephalocereus* plants have many more ribs (between sixteen and thirty) than do North American *Pilosocereus* (usually twelve or fewer). *Pilosocereus* has longish white hairs that puff out near the apex, whereas *Cephalocereus* has long beds of gray hairs on one side of the branch or tufts of cream-colored hairs that ring the branches. The fruits of *Pilosocereus* resemble large figs and are quite visible when ripe or nearly ripe; they are quite succulent and sweet. Those of *Cephalocereus* are more embedded within the woolly mats on the branches and, though edible, tend to be dry and tasteless. Human uses of these genera are few or perhaps have been abandoned by contemporary folk.

Cephalocereus columna-trajani (Karwinsky ex Pfeiffer) K. Schumann
viejito

The viejito is one of the more remarkable columnar cacti, not so much for itself (though it is unusual in its own right), but because of its quirky distribution and preferred habitat. Nearly always a single-trunked plant arched slightly toward the north, it appears to grow only in the Mexican states of Puebla and Oaxaca in confined locations. Where viejitos do grow, they often occur in great numbers—not in dense populations, but regularly spaced over entire large hillsides like so many soldiers advancing uphill. This odd appearance was the basis for the plant's name *Cereus columna-trajani*, which notes the resemblance of the cacti to columns of the Roman legions of the Emperor Trajan. The plants reach 10 m in height or slightly more. (Fig. 2.126.)

The best place to view this odd cactus in panoramic view is just southwest of Texcala, Puebla, some 12 km southwest of Tehuacán on the road to Huajuapan de León, Oaxaca. Journeying southwest from Tehuacán, one rounds a curve and is suddenly confronted by thousands of the solitary trunks stationed on a hillside. (Fig. 1.15.) The soils of the steep south-facing hillsides are calcareous, apparently a demand of the species. A few *Mitrocereus fulviceps* mingle on the lower slopes with the huge numbers of viejitos that seem to be standing watch over the ridges above. Once these hills are passed and different parent rock forms the substrate, the viejitos disappear, to be replaced, especially on lower and gentler slopes, by *Neobuxbaumia tetetzo*. On other slopes, there may be no columnar cacti at all. Individual viejitos grow here and there in other locations, but that particular hillside has the greatest known population. Other stands can be seen south of Calipan on the road to Teotitlán del Camino, Oaxaca, where large numbers are found on steep cliffs, but not in the great formations that make the Texcala site so eerily spectacular.

Apart from these specific locations, the plants have not been recorded. They join three other species of columnars found only in the Valle de Tehuacán and immediate environs: *Neobuxbaumia macrocephala*, *Pachycereus hollianus*, and *Polaskia chende*.

Not only does the viejito grow in a stooped or curved attitude, but the north side of the trunk (nearly all plants have only a single trunk) is partially to completely covered with a dense gray woolly coat resembling a beard, the pseudocephalium, hence the common name, which means "old man." Its pale pink flowers also grow exclusively on the north side, opening in April, the hottest, driest time of year in that region. Their pale color contrasts most agreeably with the gray wool of the pseudocephalium. This northern orientation of dense fibers and flowers appears to protect the flowers, which are diurnal and wilt if the sun's rays strike them. The fruits, of course, also develop on the north side. (Fig. 2.127.)

Natives of areas where the viejitos are found consider white pulp of the fruits to be edible but inferior to the various pitayo, tetecho, and xoconochtli fruits, as well as more difficult to harvest. Consequently, they are seldom

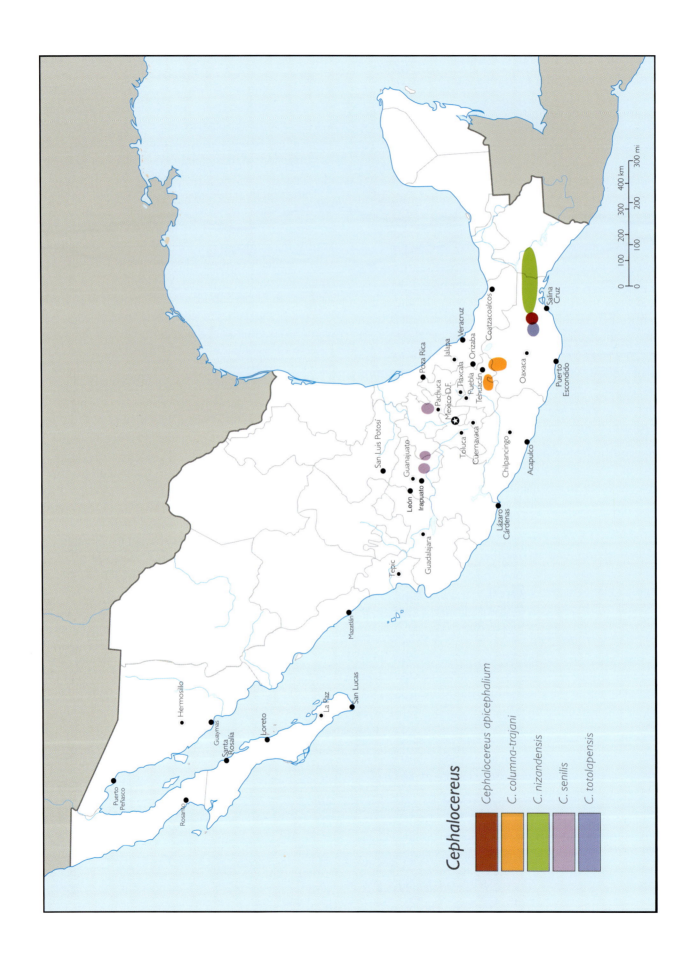

Cephalocereus

Cephalocereus apicephalium
C. columna-trajani
C. nizandensis
C. senilis
C. totolapensis

Figure 2.126. Viejitos (Cephalocereus columna-trajani) *near Texcala, Puebla.*

Figure 2.127. Viejitos (Cephalocereus columna-trajani) from above, near Zapotitlán de Salinas, Puebla.

gathered, though decades ago they were still widely eaten in the region. The cactus is not used for anything else, according to residents of Zapotitlán de Salinas, Puebla, a town adjacent to large stands of the plant. (Fig. 2.128.)

Cephalocereus senilis (Haworth) Pfeiffer
viejo, viejito

This viejo, or old man, is found on steep slopes in the profound semi-arid canyons of the Mexican states of Guanajuato, Hidalgo, and Querétaro. Its ecological requirements apparently are nearly as narrow as those of *C. columna-trajani,* for it is found only in a band around 21°N latitude between 98° and 101°W longitude at elevations below 1200 m. It appears to require less rainfall than does *C. columna-trajani,* perhaps as little as 300 mm annually. Nice stands of sentinels up to 10 m in height (reportedly reaching 15 m tall) can be found on the lower slopes of the Barranca de Metztitlán, Hidalgo, well below the elevations where the more massive *Isolatocereus dumortieri* abounds, but above the realm of *Neobuxbaumia polylopha.* These handsome plants tower above the heavily grazed thornscrub vegetation of the hillsides. From a distance, the viejos appear as silvery and often conjoined twin sentinels or descending soldiers (whereas ranks of *C. columna-trajani* appear to be *ascending!*). Unlike *C. columna-trajani,* these plants commonly branch, but usually near the ground, so that two to several long branches grow up closely parallel to each other, even touching. Above the ground, they are generally believed to branch only in response to injury. They are often greenish in youth, but with age become more grayish white with the woolly coat characteristic of the genus. However, very young cacti may appear completely covered with whitish woolly fibers. Also unlike *C. columna-trajani,* *C. senilis*'s pseudocephalium, which resembles the attaching mouth of a leech, grows on the *south* side, as do the flowers and fruits. At the plant's growing point, the newer spines tend to be white. This fresh growth, combined with the cephalium, makes the plant appear top heavy from a distance, rather like a greenish matchstick. (Fig. 2.129.)

The fruits remain on the viejo plants for some time, becoming preserved under the hot sun. They are edible, but local residents with whom I spoke were not aware of any use for the plants, preferring instead the grapelike fruits of *Myrtillocactus.* The canyon bottoms are somewhat lower and hotter than the habitat of *C. columna-trajani* in Puebla and Oaxaca, so *C. senilis* plants must endure greater heat and desiccation.

The young plants are attractive indeed. In popular surveys in the late 1930s and early 1940s, *C. senilis* was named "King of All Cacti" and the most popular cactus in the

Figure 2.128. Pseudo-Cephalium, Cephalocereus columna-trajani.

United States.[101] Collecting of young plants for export by foreigners and locals has severely compromised the long-term viability of several populations of this remarkable cactus. Although collecting has been curtailed in Mexico by threats of severe penalties, residents of the barrancas in which the cactus grows report that outsiders continue to collect it and that once vigorous and plentiful young plants have become scarce. Removable plants have become uncommon on the steep hillsides, an ominous sign for the future of the Metztitlán population. So rampant has been the plundering of the cactus that Mexican ecologists have classified the viejo as "in danger of extinction." This

Figure 2.129. Viejo (Cephalocereus senilis), Barranca de Metztitlán, Hidalgo.

classification is probably an exaggeration, for there are still huge numbers of the plant in the great canyon, and they are now watched over by three government-paid rangers. (Figs. 2.130 and 2.131.)

Cephalocereus apicephalium E. Y. Dawson

This small columnar is known primarily from the Sierra Guiengola in Oaxaca, west of Tehuantepec, where it flourishes on harsh limestone slopes, along with the odd *Neobuxbaumia tetetzo* var. *nuda*. I have seen it near Santiago Astata on the Oaxacan coast as well. The examples I found test the limits of my requirements for columnarhood, for they

barely reach 2 m tall. They are most attractive, though, with apices crowned with creamy white caps resembling a nice dollop of vanilla-dyed whipped cream atop a cactus sundae. Their habitat is rough, a substrate of treacherous limestone crags. Local knowledge of the use of the plants has escaped me. However, such strange plants must surely play a role in regional ethnobotanical lore, and horticulturally they should be most desirable. (Fig. 2.132.)

Cephalocereus totolapensis (Bravo & T. MacDougall) Buxbaum
brujo, viejito

This viejito is a single-stalked plant that reaches 11 m or more in height. It grows in tropical deciduous forests of the rugged mountains of southeastern Oaxaca, between Tehuantepec and the Valle Central, and extends well into Chiapas, where the strange creature decorates the sides of Cañon Sumidero on the Río Grijalva. On limestone canyon walls, the plants are rather common, but they are nowhere numerous and appear in only a few locations. They are difficult to photograph, primarily because they prefer steep slopes and enjoy being surrounded by rather dense dry tropical forests, a combination that usually renders them inconvenient or inaccessible to the cactus fancier. The dry season of January through April is the best time to see them in their entirety. East of Totolapan (a center of remarkable cactus diversity) on the Pan American Highway connecting Oaxaca City and Tehuantepec, they can be found on steep to sheer hillsides (usually limestone), growing along with *Escontria chiotilla, Neobuxbaumia tetetzo, Pilosocereus quadricentralis,* and *Stenocereus pruinosus,* as well as a host of burseras, leguminous trees, and many smaller cacti. (Fig. 2.133.)

Although this handsome cactus grows in a profusion of other plants of the tropical deciduous forest, it is usually easy to spot and identify, for its upper meters present rings, some including brownish white hairs, and it is capped with a thick, creamy crown of hairs. The plant often takes on a decidedly phallic appearance. The trunk has several slight constrictions at regular intervals, lending the plant a characteristic lumpy (but attractive) appearance. The flowers are pinkish, growing in a dense cluster from the apex and adding a most decorative touch to an already intriguing plant. As is the case with *C. apicephalium*, it grows in rugged, relatively remote places where knowledgeable rural folk are not often seen, hence the lack of ethnobotanical data here. One fellow who happened along while I was taking photographs acknowledged that the fruits of the *brujo* were popular with birds. Even if the fruits were tasty, they would be most difficult to harvest, for not only do the plants grow on very steep slopes, but the fruits grow

Figure 2.130.
Cephalocereus senilis,
Barranca de Metztitlán,
Hidalgo.

Figure 2.131. Young Cephalocereus senilis. *Plants may
retain the protective coat of
hairy fibers for decades.*

Figure 2.133. Brujo ("witch"),
Cephalocereus totolapensis,
*near Totolapan, Oaxaca. Other
columnar cacti here include*
Escontria chiotilla, Pilosocereus
quadricentralis, *and (probably)*
Neobuxbaumia tetetzo.

Figure 2.132.
Cephalocereus
apicephalium *growing
on limestone, Sierra
Guiengola, Oaxaca.*

Figure 2.134. Close-up of pseudocephalia of Cephalocereus totolapensis. *The silky rings remain for several years and demark each year's growth.*

Figure 2.135. Cephalocereus *cf.* nizandensis *growing on limestone inselberg, near La Venta, Isthmus of Tehuantepec.*

along the very tops of the plants. (Fig. 2.134; see also figs. 3.14 and 3.15.)

The brujos seem reluctant to branch. Most appear to have only a single stalk. Any branching is probably the result of injury or insult to the plant. I have not seen them grown horticulturally—a shame, for they are charismatic cacti.

Cephalocereus nizandensis (Bravo & T. MacDougall) Buxbaum

This attractive woolly-headed cactus grows only on nearly inaccessible limestone cliffs in the region of the Isthmus of Tehuantepec of eastern Oaxaca and probably into western Chiapas. Plants appear to reach about 3 m in height, have cute cream-colored caps similar to those of *C. apicephalium*, but are larger and stouter than the latter. Branches, when they occur, tend to grow *inward*, which gives the branched plants a characteristic shape. I have been able to approach no closer than about 15 m without risking death. I have found no local knowledge of these plants, so the odds of my being killed did not seem to justify the immense peril incurred by gaining intimate access to the cacti.

These plants appear to be capable of tolerating copious rainfall. One of the plants I photographed grows on limestone where the rainfall probably exceeds 1200 mm per year—more than 4 ft. (Figs. 2.135 and 2.136.)

Figure 2.136. Cephalocereus nizandensis *growing on limestone, Cañon el Embudo, Zanatepec, Oaxaca.*

Pilosocereus

This genus has grown and shrunk frequently in recent decades due primarily to increasing research on taxa from Brazil.[102] Zappi lists about thirty-five species including non-Brazilian taxa. Braun and Esteves disagree vehemently, listing more than forty Brazilian species, with additional taxa including subspecies and varieties.[103] Like the Brazilian taxa, the North American taxa of *Pilosocereus* are subject to ongoing scrutiny, but with somewhat greater agreement among taxonomists. The taxonomic mess of the genus in Mexico shows signs of improvement.

Pilosocereus alensis K. Weber
matagochi (Guarijío), pitaya barbona

This cactus appears in heavily forested cliffs and on steep slopes of barrancas and mountainsides in the tropical deciduous forest of Chihuahua, Sinaloa, Sonora, and apparently as far south as Guerrero, where I am not aware of having seen it. In northwest Mexico, it grows at between 500 and 1500 m elevation. The plant reaches about 6 m in height, with a few (sometimes several) lean branches. It is nowhere abundant, but it is not at all uncommon in its habitat. *Pitaya barbona* is easily distinguished from other cacti in its habitat by the eruption of cottony white woolly tufts growing toward the tips of the branches. It is the only cactus in the region with such a "beard." (Figs. 2.137 and 2.138.)

The purplish red fruits ripen in the fall, long after most other fruits have been harvested, and are prized by the Guarijíos of the sierras. Not only are the fruits succulent, sweet, and tasty, but they are also free of spines, so harvesting is simple. Many of the plants grow no more than 4 m tall, so the fruits are usually easily accessible as well, even though most plants grow on steep slopes laced with loose volcanic rock that tends to give way under the collector's feet. Unlike etchos and pitayos, however, they are not prolific producers, and they grow as widely scattered individuals, often far apart, so one cannot rely on a bumper harvest of pitaya barbona fruits. Nevertheless, they provide tasty trailside snacks for Guarijíos.

Pilosocereus chrysacanthus (F. A. C. Weber) Byles & G. D. Rowley
viejita

This species is remarkably similar to other members of the genus and but for the distribution could pass for them on superficial examination. The plants adopt a more wide-open stance than do *P. alensis* plants and are more vigorously branched. The vernacular name *viejita* means "old

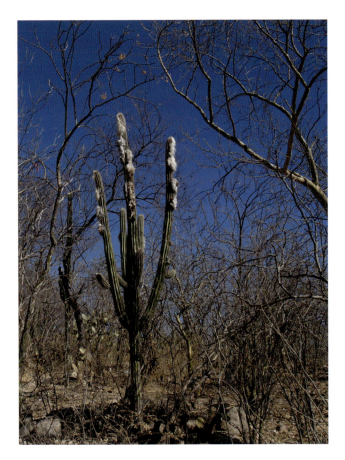

Figure 2.137. Pitaya barbona, "bearded pitaya" (Pilosocereus alensis), *on volcanic substrate, La Aduana, Sonora.*

woman," apparently because the white wool grows only toward the top of the branches, whereas the wool on *C. columna-trajani,* with which it is sympatric, covers nearly one whole side of the branch. The viejita plants grow in candelabra form up to 7 m tall, with many rather slim branches. The developing and ripe fruits and the woolly tufts can become dirty over time, bestowing on the cacti a rather ragged appearance. The cephalia develop usually on just one side of the branch, and the pink fruits emerge from this cottony mass. (Fig. 2.139.)

This species is apparently confined to Puebla and Oaxaca. I say "apparently" because I found a similar species without fruit in the state of Morelos and believe it to have been *P. chrysacanthus.* (The species of this genus can be fiendishly difficult to distinguish.) *P. chrysacanthus* is not common and is not found in thickets or forests, but it is well distributed over a wide range of elevations and is quite well known to natives. I found what I believe to have been an example at sea level near Huatulco on the coast of Oaxaca, and at nearly 2000 m elevation in the hills above Tehuacán, Puebla. Where it grows, it (like other members

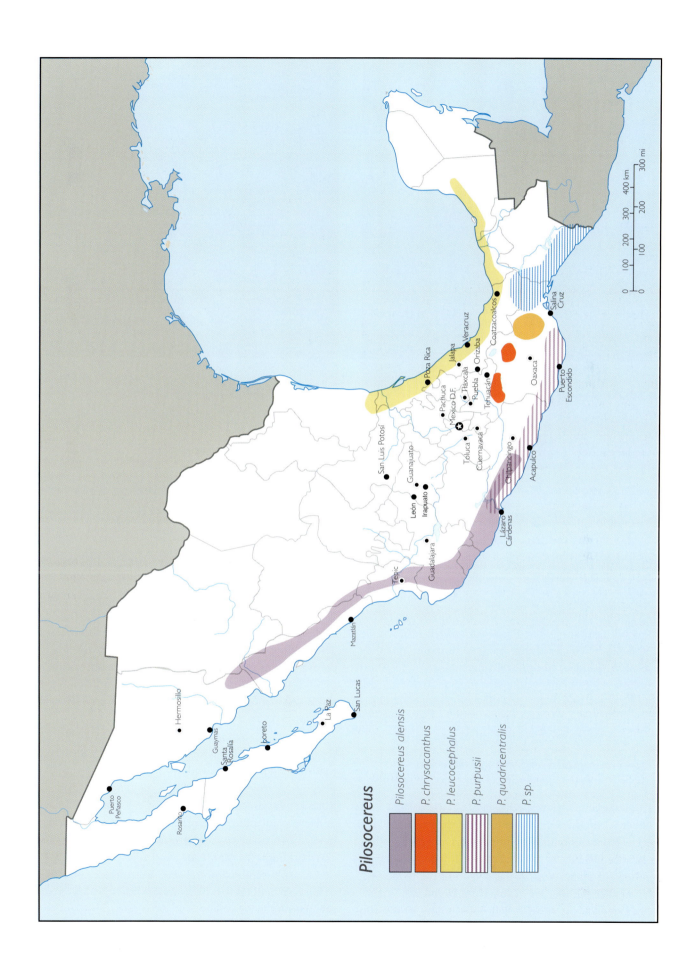

Pilosocereus

Pilosocereus alensis	*P. chrysacanthus*	*P. leucocephalus*	*P. purpusii*	*P. quadricentralis*	*P. sp.*

Puerto Peñasco
Rosario
Santa Rosalía
Guaymas
Loreto
La Paz
San Lucas
Hermosillo
Mazatlán
Tepic
Guadalajara
León
Irapuato
San Luis Potosí
Guanajuato
Toluca
Cuernavaca
Chilpancingo
Acapulco
Lázaro Cárdenas
Mexico D.F.
Pachuca
Tlaxcala
Puebla
Tehuacán
Oaxaca
Orizaba
Jalapa
Veracruz
Coatzacoalcos
Puerto Escondido
Salina Cruz
Poza Rica

300 mi
400 km
300
200
200
100
100
0
0

Figure 2.138. Pilosocereus *cf.* alensis, *Michoacán.*

Figure 2.139. Viejita, "old woman," (Pilosocereus chrysacanthus), *near Reyes Metzontla, Puebla. Residents of the region esteem the fruits of this cactus.*

of the genus) is easy to spot, even in the rainy months, due to the white tufts at the tips of the branches.

The fruit husks split open to reveal the bright pink-purple pulp, which is edible, though dry, sweet, and full of grainy but tasty seeds. The husks are covered with woolly white hairs in which various airborne contaminants may become embedded, so the consumer must pay heed not to pollute the pulp and perhaps to wash it. The fruits are often gathered—easily, because they are spineless—but for domestic consumption only, for they do not appear to be sufficiently attractive, abundant, or juicy to market.

Pilosocereus quadricentralis E. Y. Dawson
viejito

In this large genus, a few species stand out as excellent fruit producers, and this viejito is one. In its favorite habitat, the low, rugged mountains of southeastern Oaxaca, it is common indeed. On rocky hillsides east of Totolapan, Oaxaca, hundreds of plants seem to beckon from the landscape. A far smaller number are visible from Highway 175 near the bridge over the Río Guelatao, a tributary of the Río

Grande Quiotepec. The abundant strands of cotton that cover the upper branches cause the branches to stand out like myriad points of light in the forest (notably in the dry season, November to May). (Fig. 2.140.)

The plants are tall for the genus, reaching 8 m in height. The height is important for survival in the dry tropical forest of the mountains of southeastern Oaxaca because large plants lacking sufficient height will be cut off from sunlight during four months of the year. For cacti, especially, this lack could prove a mortal deficiency.

The fruits of this viejito are superb in color, texture, and flavor, and are readily accessible on the upper surfaces of the branches. The husks are thin and spineless (as are all *Pilosocereus* fruits), and the juicy, dark reddish-purple pulp is easy to expose and even easier to eat. The pulp is sweet and juicy. I introduced an urban Oaxacan to the fruits, and he was astonished that he had never before been offered one. Although the plants are not prolific fruit producers, they are sufficiently common, and the fruits are easy enough to harvest that they should be more widely exploited commercially. (Fig. 2.141.)

Figure 2.140. Viejito (Pilosocereus quadricentralis) near Totolapan, Oaxaca.

Pilosocereus purpusii (Britton & Rose) Byles & G. D. Rowley
viejo, cabeza de viejo

P. leucocephalus (Poselger) Byles & G. D. Rowley
cabeza de viejo

These two viejos grow on opposite sides of the Mexican subcontinent. Both produce edible fruits, but occur in rather lush habitats where their potential virtues tend to go unnoticed. *P. purpusii* grows along the Pacific Coast in the states of Guerrero and Oaxaca, whereas *P. leucocephalus* is found from Tamaulipas on Mexico's Atlantic Coast south into Central America. (Figs. 2.142 and 2.143.)

Both species produce edible fruits, but grow in tropical conditions where other fruits and wild foods abound, so they are not renowned for their virtues. However, *P. leucocephalus* is familiar to residents of the canyon of the Río Actopan, Veracruz. It is a much larger and taller plant than *P. purpusii*, and the white tufts at the tips of spreading arms are a common sight. Residents report that they have heard of people eating the fruits, but they themselves have not. This is understandable because the area abounds with

Figure 2.141. Fruit of Pilosocereus quadricentralis. *The husks have no spines, so the fruits are ready to eat when picked. Although very tasty, the fruits are not as abundant as those of most species of* Stenocereus.

Figure 2.142. Cephalia of Pilosocereus leucocephalus, *Río Actopan, Veracruz.*

Figure 2.143. Pilosocereus purpusii, *Cañon Infiernillo, Michoacán.*

Figure 2.144. Pilosocereus lanuginosus, *Bonaire, Netherlands Antilles.*

the prolific fruit producer *Stenocereus griseus* and many thousands of mango trees. *P. purpusii* is easily overlooked in the thick tropical deciduous forest of the Oaxaca and Guerrero coast, for it seldom exceeds 3 m in height and is a rather skinny, unpretentious cactus competing with large *Neobuxbaumias, Pachycerei,* and *Stenocerei* for attention. I have not seen its fruits, but they should be edible and easily harvested.

Pilosocereus lanuginosus (L.) Byles and G. D. Rowley
dadu

This tall but trunkless giant is common and prominent on Bonaire and Aruba, where it reaches 10 m or more in height, with wildly flailing branches. It may be the same as *P. tweedyanus* of Ecuador, but its habit in the Antilles is utterly different. In the Antilles, it is more handsome and taller, and it has thicker branches. Anderson refers tweedyanus to *P. lanuginosus,*[104] but the two appear rather different to me. The fruits of *lanuginosus* are widely eaten, but not numerous. This cactus is also planted from time to time as part of living fences and seems to grow quickly. It is also common in the arid valleys of northwestern Venezuela. (Fig. 2.144.)

Figure 2.145. Pilosocereus *sp., Cintalapa, Chiapas.*

Figure 2.146. Fruit of unidentified Pilosocereus *sp., near Cintalapa, Chiapas. Note that birds have already begun eating the fruit.*

Pilosocereus sp.
cabeza de viejo

This tall plant, reaching 8 m in height, possibly more, is common in the more moist forest of the eastern Isthmus of Tehuantepec well into Chiapas, where I found it growing in tropical deciduous forest near the city of Cintalapa. The rinds of the fruits are wrinkled and ugly, so the fruits are usually passed over for those with a more palatable appearance. The rib count and spine arrangement do not correspond to the description of any *Pilosocereus* species I have encountered. It may be a race of a taxon already described. I await others' work on its classification. (Figs. 2.145 and 2.146.)

The Goblet Cacti: Jiotillas, Garambullos, Chendes, and Chichipes

Members of this group of six Mexican species exhibit strong evolutionary affinities and a marked tendency to grow into goblet or candelabra form. The plants have many branches for the most part, the arms usually ending in a vertical alignment. They are not gigantic or particularly tall (the tallest probably does not exceed 10 m in height). But they are a delight to behold. Healthy individuals growing under the right circumstances exhibit a most agreeable symmetry and a flurry of arms, all arching heavenwards. In the village of Reyes Metzontla, Puebla, five of the six species described here can be found, at least four of them growing together in the same yards. All have edible fruits, and the fruits of at least four are gathered and sold in local markets. Cornejo and Simpson suggest that all species may represent a single genus.[105] However, the fruits of the existing genera differ so much that referring all to one genus seems unwise. *Myrtillocactus eichlamii* Britton & Rose is reported from Guatemala. I have not seen it.

Escontria chiotilla (Weber) Rose
jiotilla, quiotilla, xonochtli, *xuego* (Oaxaca)

Near Presa Infernillo on the Río Balsas, Michoacán, this many-branched cactus of the tropical deciduous forest

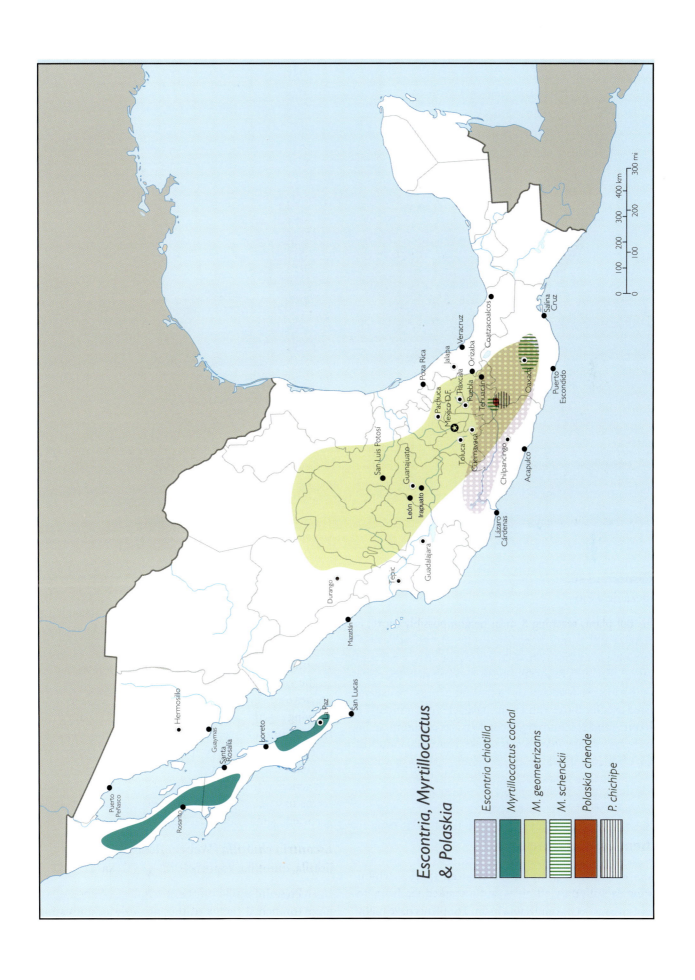

Escontria, Myrtillocactus
& Polaskia

Escontria chiotilla
Myrtillocactus cochal
M. geometrizans
M. schenckii
Polaskia chende
P. chichipe

Puerto Peñasco
Rosario
Hermosillo
Guaymas
Santa Rosalía
Loreto
La Paz
San Lucas
Durango
Mazatlán
Tepic
Guadalajara
León
Irapuato
Guanajuato
San Luis Potosí
Pachuca
Mexico D.F.
Toluca
Cuernavaca
Chilpancingo
Lázaro Cárdenas
Acapulco
Puerto Escondido
Oaxaca
Tehuacán
Puebla
Tlaxcala
Orizaba
Veracruz
Jalapa
Poza Rica
Coatzacoalcos
Salina Cruz

0   100 200 300 400 km
0   100 200 300 mi

Figure 2.147. Jiotilla (Escontria chiotilla), *Valle de Tehuacán, Puebla.*

(or former tropical deciduous forest) grows in extensive thickets known as *quiotillales*. It is also found in large numbers in canyons and valleys above 1000 m elevation across a broad band of south central Mexico in the states of Guerrero, Oaxaca, and Puebla. In this semiarid region, the cactus is common, but does not grow in the density found in the forestlike stands around Presa Infiernillo. In the rolling country of north-central Oaxaca and well into southwestern Puebla, it is a very common cactus. Near Totolapan, Oaxaca, numerous large plants grow to the exclusion of most other species of columnar cacti, and its fruits are routinely harvested commercially. (Fig. 2.147.)

Identification of jiotillas from a distance is usually simple, for the strongly angular plant often resembles an upside-down triangle or pyramid and its ribs are pronounced and *without* the large, wartlike areoles characteristic of *Myrtillocactus*. It has numerous smaller areoles that sometimes almost merge into each other on the ribs.

Near the basal portion (bottom) of each is a tearlike extension with a spine growing out. The plant usually has no trunk or only a diminutive one, much less so than either *Myrtillocactus* or *Polaskia*, both of which it superficially resembles. The margin (edge) of the ribs is usually straight, not knobbed as is the case in *Myrtillocactus*. The many-branched trees grow up to 6 m in height, occasionally taller, especially around Totolapan. The fruits grow at the apex of the branches, whereas in *Myrtillocactus* they grow along the sides.

In May and June, the branch tips of jiotillas produce many delicate bright yellow flowers, usually 3–4 cm wide, the only columnar cactus in the region with yellow blooms. The small brownish-red fruits ripen in late June into July and are popular; they can often be seen for sale in markets in southern Mexico. I have seen them for sale in Acatlán and Tehuacán, Puebla, and near Totolapan, Oaxaca. They differ from other cactus fruits in that they are covered with

Figure 2.148. Jiotilla in fruit, Nochtixtlán, Oaxaca.

Figure 2.149. Xuegos (*fruits of* Escontria chiotilla), *La Reforma, Oaxaca.*

large scales rather than spines, so the ping-pong-ball-size fruits resemble fir cones more than cactus fruits. To eat them, one simply squeezes the scaly husk, and the sweetish pulp pops out and can be sucked into the mouth. The fruit pulp is sweet and has a delicate flavor. The seeds are large, 1.5 mm wide. (Fig. 2.148.)

The plants have an affinity for overgrazed and poor or disturbed soils and seem to flourish both on hillsides and on flats. On rocky, eroded hillsides and on trampled bottomland, they still produce numerous fruits. Given this remarkable flexibility, they demonstrate a strong potential for commercial fruit production in areas otherwise bereft of economic productivity.[106] To my knowledge they have not been commercially cultivated, although I have seen reports that orchards have been planted in eastern Oaxaca. *Escontria*'s wide range of distribution indicates that it is a hardy cactus, indeed. (Fig. 2.149.)

Escontria chiotilla appears to be evolutionarily related to proto-*Stenocereus* and to be a less specialized progenitor of the genera *Myrtillocactus* and *Polaskia,* which share its inverted triangle shape.[107] In other words, it is linked to the ancestors of the other genera. Its abundance and widespread distribution in southern Mexico also suggest its early status among columnars.

Escontria chiotilla is a most versatile cactus, tied with *Stenocereus stellatus* for the honor of the southern Mexican columnar cactus with the most uses. Casas and Barbera note seven types of use: edible fruits, livestock fodder,

fermentation into alcoholic beverage, edible seeds, edible stems and flowers, living fences, and fuel wood.[108] In *Escontria* territory, La Michoacana, a popular ice cream chain, frequently sells *paletas* (popsicles) of the pulp mixed with water and sugar.

Myrtillocactus geometrizans (Martius) Console
garambullo, padre nuestro, *tepepoa nochtli* (Nahua), blue candle cactus, blue myrtle cactus

M. schenckii (Purpus) Britton & Rose
garambullo, vichisova

M. cochal (Orcutt) Britton & Rose
cochal

We might well call these three species the "berry" cacti, as their generic Latin name suggests (*myrtillo* means "berry"), for their fruits resemble large grapes or currants. Because the fruits of *M. geometrizans* have neither spines nor glochids, they can be plucked from the tree and eaten. *Myrtillocacti* also can be distinguished from the cacti of other genera because they alone (except for *Pachycereus [Lophocereus] schottii*) produce several flowers (and fruits)—between two and twelve—from each areole. When the trees are in flower, their branches can appear covered with flowers, like no other cactus. (Figs. 2.150 and 2.151.)

M. geometrizans, the garambullo or blue candle cactus, is common and well known in central and southern Mexico. Individual plants may become giants, exceeding

Figure 2.150. Flowers of Myrtillocactus geometrizans.

Figure 2.151. Flowers and fruits of Myrtillocactus schenckii, *Reyes Metzontla, Puebla.*

7 m tall and equally wide. This handsome cactus is one of Mexico's most broadly distributed columnars, ranging from Tamaulipas southwest to Durango and south to Oaxaca. In parts of central Mexico, especially in the arid high plains of the states of Hidalgo, Querétaro, and San Luis Potosí, hundreds of thousands of these candelabra-like cacti cover the flat landscape, offering the impression of a forest of arms, at times the only large plants to be found. When I last saw such a forest in 2000, though, many plants near the highway were dead or dying, casting an eerie mood over the panorama. Farther away from the roadway the plants seemed healthier, suggesting that vehicle exhaust (or accumulated lead?) may be taking a toll on those hapless individuals trapped near the highways. (Fig. 2.152.)

M. geometrizans is readily distinguished by its spreading, candelabra-like habit and many arms, each with five to six ribs lined with raised areoles 1–4 cm apart, which from a distance resemble warts spaced along the ridges. The plant often has more than one trunk from which branches divagate like ascending family trees, so that the crown is made up of many dozens of branch tips of roughly equal height. This manner of growth gives the overall shape of

an inverted cone with beveled edges, but usually having a more distinct trunk than is the case with *Escontria*. In an area often devoid of shade trees, the conical shape produces dense, cool shade during the heat of the day when the sun is directly overhead. At times, the plants are wider at the crown than they are tall. A subspecies with more prominent areoles, var. *grandiareolatus,* is common throughout the Valles de Tehuacán and Zapotitlán. Their areoles are so raised that they resemble the note buttons of an accordion. (Figs. 2.153 and 2.154.)

New branches of *M. geometrizans* are an attractive bluish green (hence the English name), and the plant produces abundant greenish white, 3-cm-wide flowers. The fruits of garambullos mature in late spring well into summer. They are small in diameter, 1–2 cm, grapelike, purplish to dark blue, and without spines. These tasty fruits are said to be sweetest when they are somewhat dry and wrinkled like big raisins. They are popular, indeed, eaten off the plants as a snack or steeped into a refreshing beverage. In the Valle de Tehuacán, they are widely marketed as a fruit and are made into jam, but are more commonly used to make an alcoholic beverage called *vino de garambullo.* In

Figure 2.152. Garambullo (Myrtillocactus geometrizans), *Sierra Gorda, Querétaro.*

Figure 2.153. Garambullo near Ejutla, Oaxaca. Note the dramatically different habit from that in fig. 2.150.

Figure 2.154. Garambullo (Myrtillocactus geometrizans) and baboso (Pachycereus hollianus), Zinacatepec, Puebla.

Figure 2.155. Fruits of garambullo. Note multiple fruits from a single areole, right of center.

Zapotitlán de Salinas, I tried it and found it alcoholic and sweet, reminiscent of cough medicine, perhaps in need of an experienced enologist's skills. (Fig. 2.155.)

In the semiarid hills and the vast, sweeping plateaus of Querétaro and San Luis Potosí north of Mexico City, where cactus fruits are abundant from May through October, the fruits of the garambullo (or the *padre nuestro,* as it is widely called in the northern part of its distribution) are harvested, dried, and sold as raisins. They are also marketed commercially in jamlike preserves and even mixed with sugar water, frozen, and widely sold as popsicles. Countless tons of the fruits ripen each year, far more than any small group of individuals can harvest.

The cactus prefers arid mesquite, catclaw, or yucca *bosques* and flats, but it will also grow on hillsides. When it finds a site to its liking, it reproduces in great numbers that grow thickly, indeed, sometimes in intimate association with other columnars. In parts of the Valle de Tehuacán, especially where the intimidating catclaw *Mimosa luisana* grows thickly, recruitment of *M. geometrizans* var. *grandiareolatus* seedlings appears to be successful, and the

overall population seems to be thriving. A wide variety of sizes of the pretty bluish-green juveniles are in evidence. Buttons of the seedlings spring peyotelike from the soil. Where goats and machete-wielding campesinos manage to keep the catclaw at bay, neither garambullo nor any other columnar cactus appears to be reproducing. And judging from the enormous numbers of the plants in the plains of central and northern San Luis Potosí state, some strains must be able to tolerate temperatures near or even below freezing. In those vast expanses, only a local yucca matches the cactus in size.

In Puebla and Oaxaca, where fuel wood supplies are chronically overharvested and many thousands of households cook over fires, the wood of the garambullo, which is extremely hard for a cactus, is commonly exploited as firewood, always an ominous sign for the well-being of a cactus species. According to local experts, the wood of the garambullo arms is durable and springy enough to weave into walls and ceilings, which are then covered with mud or other waterproofing.

A closely related garambullo, *M. schenckii,* grows in the eastern portions of the wide, semiarid central valleys of Oaxaca and the foothills below. The plant also is common near the town of Reyes Metzontla, Puebla, and crops up here and there in the southern Valle de Tehuacán. I have heard reports of a disjunct population in Hidalgo as well. The juveniles of this species are more truly green (even yellowish) than the young *M. geometrizans* plants, which are nearly a turquoise color, and the *schenckii* spines are yellowish. Adults are more difficult to distinguish, but *M. schenckii* has at least twice as many prominent areoles on the ribs as *M. geometrizans,* usually less than 1 cm apart. The *schenckii* flowers have yellow anthers; its fruits are more noticeably reddish and bear tiny spines, unlike the spineless fruits of *M. geometrizans. M. schenckii* seems to be especially prominent on pre-Columbian archaeological sites such as Mitla and Yagul. Guides at these locations snack on the fruits. Some of them have estimated that the plants live for at least one hundred years. (Figs. 2.156 and 2.157.)

Very large individual *M. schenckii* plants grow in the rugged hill country east of the central valleys of Oaxaca. These handsome cacti reach 9–10 m in height and grow equally wide, with hundreds of arms. When I showed photographs to cactus experts, they were incredulous that this species could have so many arms. The plants are often too tall for unassisted humans to reach the fruits, but their shade is a boon to livestock and sweaty humans during the hot months of April and May. (Figs. 2.158 and 2.159.)

In Baja California, the more diminutive (or at least shorter) *cochal (M. cochal)* grows widespread, reaching

Figure 2.156. Myrtillocactus schenckii, *Yagul, Oaxaca.*

Figure 2.157. Myrtillocactus schenckii, *Mitla, Oaxaca.*

Figure 2.158. Fruits of Myrtillocactus schenckii, *La Reforma, Oaxaca.*

Figure 2.159. Myrtillocactus schenckii *near La Reforma, Oaxaca.*

nearly to the U.S. border. Though found mostly in the northern third of the peninsula, a disjunct population also grows in the Cape Region and seems to flourish there in the greater rainfall.

These handsome but somewhat rangy plants can hardly be confused with any other. Cochales have fewer branches and are rather shorter (usually less than 4 m tall) and less spreading than other members of the genus. Their branches may grow more densely, however. The areoles on the branches are prominent and noticeable from a distance, and around the edge of each areole emerge five potent, stout spines and a single longer spine from the center. All the cochal spines are far longer than those of the other two *Myrtillocactus* species. The central spine may be nearly 2.5 cm long. They are greenish yellow in color, tending to turn even yellower during droughts. The plants bloom throughout the year. They are to be found in a wide variety of desert habitats, extending much of the length of the peninsula to within a hundred kilometers of the California border. The reddish fruits are eaten and were apparently an important source of food for pre-Columbian peoples.

Cochal is a lowland plant, often growing on hillsides above the Pacific Ocean, whereas the garambullos prefer highlands. How the cochal or its progenitors came to grow in Baja California without intermediate plants found for more than 500 km is a mystery. Perhaps they evolved from *M. geometrizans*, a population remaining when Baja California split off from the mainland some 5–6 million years ago, and found their best niche in the northern portion of the Baja California peninsula. (Fig. 2.160.)

Polaskia chichipe (Gosselin) Backeberg
chichipe

P. chende (Gosselin) Gibson & Horak
chende

These two cacti of highly restricted range join jiotillas and the cochal in bearing a species name that is also their common indigenous name. These worthy plants are confined to a small portion of southwestern Puebla and a tiny portion of northern Oaxaca. On the basis of data I have gathered, it appears that *Polaskia* has the most limited range of any columnar cactus genus, far narrower than the range of the other goblet cacti and narrower even than the ranges of the very restricted *Backebergia* and *Mitrocereus*. Chichipes are rather more common than chendes, which are endemic to a very narrow region—little more than 500 square km—and are uncommon even within that region.

Figure 2.160. Cochal (Myrtillocactus cochal), *Sierra San Francisco, Baja California.*

I suspect that it would not be too difficult to count all the wild chende plants in the world. They must have very strict and narrow demands to be confined to such a tiny area of the globe. Or perhaps they are so recently evolved that they have not yet had time to spread into the surrounding landscape. Chichipes, however, can be found on hilltops and gentle hillsides over an area at least 80 km wide, from southwestern Puebla to villages south of Huajuapan de León, Oaxaca, where they are also well known to local people. Most chichipe plants grow at an altitude greater than 1500 m elevation.

The best place to study both species is in and around Reyes Metzontla, Puebla, a Popoloca village in an outlying side drainage of the Valle de Zapotitlán. Chichipes are common in the town and surrounding hills, and a few chendes grow in yards, along with several species of *Stenocereus*, numerous *Escontria* plants, and an occasional *Pilosocereus*. Wild chendes grow on the higher slopes at some distance

from the village, as they do on the outskirts of the nearby town of Acatepec, Puebla. Their preferences for slopes and ridge tops makes them quite inaccessible to anyone unwilling to hike considerable distances to view the plants. (Figs. 2.161, 2.162, 2.163.)

Superficially the two species are difficult to distinguish. Both have many branches and are candelabra-like in appearance. However, chendes have fewer ribs (six to eight, compared with eight to twelve for chichipes). Chende flowers are solitary and open in the morning, whereas chichipe flowers are grouped at the apex of branches and open at night. Chende flowers are also considerably larger than chichipe flowers, and the new branches of chendes tend to be considerably longer than those of chichipes (2 m as opposed to 1 m in length), lending the chende plants a more rangy appearance. Chende fruits have dark purple-red pulp, whereas chichipe fruits have strawberry red pulp. Chichipe fruits are handily harvested from the tips of the branches, but chende

*Figure 2.161.
Chichipe* (Polaskia
chichipe), *Reyes
Metzontla, Puebla.
Note flowers and
fruits at the apex of
the branches.*

*Figure 2.162.
Chichipe flower, Reyes
Metzontla, Puebla.*

Figure 2.163. Ripening chichipe fruits, Reyes Metzontla, Puebla.

fruits grow all over the branches and may be more difficult to gather. Even knowing these differences, though, I found I could walk casually among plants of the two species when they are not in flower and fail to observe any distinction. Natives of the region where the plants grow can distinguish them instantly. (Figs. 2.164, 2.165, and 2.166.)

Faced with the plants' similarity, I have decided that the easiest way to tell them apart (outside of the fruiting season of August) is simply to enlist the help of villagers. To my delight, when I asked a housewife where I might find a chende, she pointed to one growing in her yard. It and a chichipe grew closely together. Seeing them side by side was educational, but not sufficient for me to make discriminating between them automatic.

The fruits of both of these handsome plants are highly edible. Chende fruits are somewhat larger, the size of golf balls. These fruits are often mashed vigorously and then mixed with water to make a *refresco,* or soft drink, with a most agreeable color. The color is so intense that it is used to produce a purplish dye for tinting wool that is woven in

the region, though I haven't seen the dye made or applied to the wool. It is also used to give a purplish tint to beverages as well, perhaps an accomplishment in deceitful advertising.[109] Chichipe fruits (called *chichitunas,* an indigenous name) are eaten and marketed locally, but I have not found similar accolades about their uses. A regional salsa is prepared from them.

Residents of Reyes Metzontla raise plants of both species from cuttings, but most of the plants in the village are wild, or, more accurately, managed in situ.[110] Many villagers built their houses where wild specimens (especially chichipes) could be found conveniently nearby, incorporating them inside their compound with a simple fence, thus creating a personal orchard of plants they manage for production. A chende I photographed grew in a yard also occupied by chichipes, jiotiollas, malinches, pitayos de mayo, pitayos de octubre, and xoconochtlis. I rather suspect that it was planted long ago as a cutting. An elderly Popoloca woman who believed she had planted it as a child also tended another large chende.

Figure 2.164. Chende (Polaskia chende), Acatepec, Puebla. This plant has undoubtedly been grown from a cutting.

Figure 2.165. Wild chende, Reyes Metzontla, Puebla.

Figure 2.166. Chende flower.

The Mexican Loners: *Isolatocereus, Mitrocereus,* and *Backebergia*

Some cactus authorities (but by no means all) currently assign each of the massive cacti *Isolatocereus dumortieri, Mitrocereus fulviceps,* and *Backebergia militaris* to a monotypic genus. *Isolatocereus* has been split away from *Stenocereus* into its own genus. *Mitrocereus* has recently been referred to *Pachycereus*. Some taxonomists have also referred *Backebergia militaris,* which through much of its botanical history occupied its own genus due to its highly individualistic appearance and limited distribution, to *Pachycereus*. Genetic affinities aside, out of pure affection for their distinctive characteristics I view these three as so sufficiently different from other genera and as solitary members of their own genera that they must be cast as loners. If lumpers have their way, the three genera will cease to exist.

Though their habitats extend to within 10 km or so of each other, *Isolatocereus* and *Mitrocereus* do not appear to overlap, and *Mitrocereus* and *Backebergia* grow even more distantly, never closer than 150 km, from my reckoning. At least a few specimens of *Isolatocereus* grow a few kilometers southwest of Acatepec, Puebla, and a few individuals of *Mitrocereus* are to be found no more than 15 km to the northeast. Of the two, *Isolatocereus* is by far the more common, growing in rather dense populations over parts of its considerable range (especially the western upper hillsides of the Barranca de Metztitlán), whereas *Mitrocereus* never appears to grow in densities of more than a scattering of individuals on a handful of hillsides. *Backebergia* and *Isolatocereus* may commingle, but I have not seen them do so. *Backebergia*'s habitat is apparently too low in elevation for *Isolatocereus*.

Isolatocereus dumortieri (Scheidw.) Backeberg
malayo, órgano

This candelabra-like giant (up to 13 m tall and nearly as wide), called *malayo* in Puebla and Oaxaca, grows in a variety of habitats in central Mexico, from the profound canyons of Hidalgo and Querétaro south to the semiarid hills of Puebla and Oaxaca, and on to Cañon de Zopilote of the Río Balsas in Guerrero and the steep slopes of La Cañada in northwestern Oaxaca, which drains into the Atlantic. The plant invariably has a large to massive trunk with upright branches, sometimes affording it a striking resemblance to *Pachycereus weberi* (the chico) from a distance. The ribs of the branches are few, usually five or six, whereas *P. weberi* has at least eight. *I. dumortieri* is light green in color; *P. weberi* is bluish green. The areolas of the former are numerous, less than 1 cm apart, whereas those of the latter are at least 3 cm apart. Spines of the malayo are relatively short and grow in small numbers, but chicos are well armed with longer spines. These characteristics combine to give the malayos a clean appearance, even though individual cacti may sport rotten or dead arms, especially the extrawide plants found in the canyon of the Río Acatlán. The flowers are nearly tubular, which renders the plants distinct from *Stenocereus,* under which they were classified for years.

Forests of these órganos (as they are called in the northern part of their range) dominate the hillsides of Hidalgo's Barranca de Metztitlán, and smaller numbers of truly massive plants appear in the drainage of the Cañon de Acatlán in Puebla. The plants of the more northern states tend to be less candelabra-like; that is, the branches tend to grow more closely to the central axis and parallel to it, whereas branches of the southern plants grow horizontally for some distance before growing vertically, thus rendering a candelabra-like appearance to the plants there. I suspect that different taxa are involved in the two populations, but I have only superficial evidence for this theory. (Fig. 2.167.)

The malayos of Puebla and Oaxaca grow nearly (but not quite) as large as *P. weberi*. I have not found the two species growing in the same habitat, even though they may be found within a few kilometers of one another. The malayo appears to prefer elevations higher than 1500 m, and *P. weberi* prefers lower elevations. South of Huajuapan de León the malayo appears in the higher Valle de Acatlán. It populates a narrow elevational band, then disappears. Downstream, *P. weberi* appears abruptly. (Fig. 2.168.)

The malayo's purplish buds and pink to white flowers appear at the apex of the branches and are thus often difficult to see. The bright red to reddish-orange (sometimes translucent) fruits, though frustratingly inaccessible, are sweet and tasty and usually considered worth the effort of harvesting. They are more popular in the northern states, where the variety of available cactus fruits is considerably less than in Guerrero, Oaxaca, and Puebla. The wood is flimsy and splintery, rendering it less than desirable as lumber or fuel wood. This characteristic seems odd, given the ponderous weight that the structure must bear. I once examined a dead branch closely and found I could easily break it across my knee, whereby it splintered into pieces no stronger than cane.

For decades, malayo was assigned to *Stenocereus*. Backeberg originally proposed the genus *Isolatocereus,* but the classification was suspect because of his penchant for splitting and creating whole new genera where none was appropriate. Gibson, however, observed that the malayo

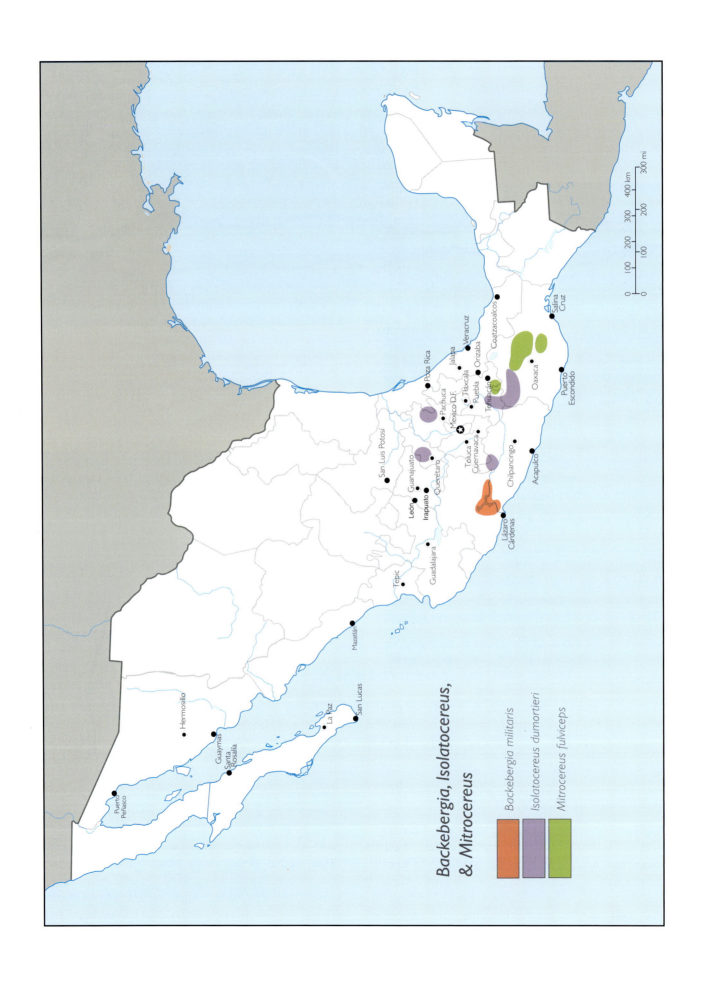

Backebergia, Isolatocereus,
& Mitrocereus

Backebergia militaris

Isolatocereus dumortieri

Mitrocereus fulviceps

Figure 2.167. Isolatocereus dumortieri, *Barranca de Metztitlán, Hidalgo.*

plant lacks silica bodies in the stem cells, unlike other unequivocal members of *Stenocereus,* so he referred it to its own genus,[111] and Backeberg's taxonomy regained legitimacy through the back door. Hunt, Taylor, and Charles have re-referred it to *Stenocereus.*[112]

The malayo is a remarkable cactus worthy of greater publicity for both its immensity and its stateliness. The forests of these huge plants covering the upper slopes of Barranca de Metztitlán deserve protection.

Mitrocereus fulviceps (Weber) Backeberg ex Bravo
órgano de cabeza amarilla

This massive, majestic cactus has narrow ecological requirements and is one of the rarer large columnars. It differs enough from other great cacti and is sufficiently uncommon to have suffered the indignity of frequent attacks by botanists on its taxonomic status, being kicked back and forth among various genera. The question of its correct location among cacti is by no means settled, but for the time being the sole member of *Mitrocereus* will do. Anderson, reflecting the findings of the International Consensus on Cacti, assigns it to *Pachycereus,* but I remain unconvinced.[113] (Fig. 2.169.)

The cactus appears to be limited to scattered populations in semiarid canyons between 1500 and 1800 m elevations in Puebla and Oaxaca. A robust population grows in Puebla southwest of Tehuacán on south-facing slopes near Texcala, on the hills above Zapotitlán de Salinas, and a few kilometers east of Santiago Acatepec near Reyes Metzontla. It seems quite happy growing amidst groves of *Cephalocereus columna-trajani* in the Valle de Zapotitlán and with *Neobuxbaumia macrocephala* near Reyes Metzontla in the upper reaches of that valley. Another population grows some 150 km to the south at the southern end of the semiarid valley called La Cañada in northeastern Oaxaca. It is the lone species of columnar cacti on the steep north-

Figure 2.168. Malayo (Isolatocereus dumortieri), *Valle de Acatlán, Oaxaca.*

Figure 2.169. Mitrocereus fulviceps *in habitat, La Cañada, Oaxaca.*

Figure 2.170. Mitrocereus fulviceps, *Zapotitlán de Salinas, Puebla. Author is in the center of the photo.*

Figure 2.171. Mitrocereus fulviceps *in fruit, Reyes Metzontla, Puebla.*

facing valley slopes above La Cañada at an elevation well above 2000 m, where it grows in the company of several large varieties of oak (*Quercus* spp.). The La Cañada plants stand out strikingly from the scrubby chaparral that covers the slopes. A separate population, apparently a smaller number of plants, grows on steep slopes above the Río Guelatao north of Oaxaca City, only a few kilometers from the town of the same name (Guelatao), where Benito Juárez was born. These populations are all in watersheds draining into the Gulf of Mexico. Graham Charles reports *Mitrocereus* from San José Lachiquiri, Oaxaca, in a canyon draining into the Río Tehuantepec, which in turn drains into the Pacific.[114] (Fig. 2.170)

As is the case with *N. macrocephala,* the number of *Mitrocereus* is surely limited to no more than a few thousand individuals. The Oaxaca populations appear to be thriving, and recruitment of small plants is encouraging. The cacti are easy to spot from great distances due to their color and massive size. It seems odd that a great cactus with

such limited distribution should tolerate such a variety of altitudes and plant associations. (Fig. 2.171.)

The common Spanish name, *órgano de cabeza amarilla,* which translates "yellow-headed organ pipe," derives from the yellowish-orange crown on the branches produced by a dense woolly covering that is accentuated during flowering season. The buds appear at the apices of the branches and are covered with a dense pile of yellowish, felty mat. The white flowers open at night and are apparently pollinated by bats. This cactus may be one of the several species of columnars completely dependent on bats for pollination. (Fig. 2.172.)

Mitrocereus might be confused with the similar-appearing and almost equally rare *Neobuxbaumia macrocephala,* for both have woolly caps (I confused them for a time). *N. macrocephala*'s cap tends to be reddish, however (its common name is *órgano de cabeza roja,* "red-headed organ pipe"), its branches grow more compactly and close to the main axis, and it has many more ribs, seventeen to twenty-five compared to *Mitrocereus*'s eleven to fourteen. When the two are seen growing side by side (not a frequent occurrence), the yellow-green color of *Mitrocereus* becomes obvious, readily distinguishing it from the nondescript dusty green of *N. macrocephala*. In fact, the *Mitrocereus* branches tend toward yellow on the spectrum, and the species can often be distinguished by color alone. (See fig. 3.2 for a photo of these two species together.)

I have not had an opportunity to taste the ripe fruits. Those I sampled were old, dry, and unappetizing in appearance. Local experts have explained that they do not eat the fruits because unlike so many other cactus fruits that ripen at the same time, they are not tasty and sweet. One fellow said they were edible, but not very tasty. Where *Mitrocereus* grows, plantations of *Agave salmiana* are common, and the pineapple-like juice *aguamiel* from that plant is regularly harvested in great quantities and naturally ferments into pulque. No one would want to subsist on dry, mealy fruits when nectar of the mother liquid of the intoxicating pulque is readily available. Harvesting fruits of such questionable virtue is an onerous proposition, anyway, for one would have to gain access to the woolly tips of the órgano's branches and grope inside the cephalium to extract the fruits. The plants always grow on steep slopes, so fruit gathering would require stepladders, considerable expertise, and, for the prudent, generous personal insurance policies. One could starve to death gathering these fruits.

I have been unsuccessful in determining other uses for the plant, its flowers, and its fruits. The fact that it is not used may be a reflection of the relative inaccessibility of the fruits (at the tips of the lofty branches) and the plant's scarcity, especially in a land in which the luscious fruits of other cacti and cultivated orchards of tropical fruits are abundant and readily accessible.

Backebergia militaris (Audot) Bravo ex Sánchez-Mej.

tiponchi, grenadier's cap

This handsome, spectacular plant[115] is found only in the hot lowlands of Michoacán and Guerrero, especially in the vicinity of Presa Infiernillo, a hot spot of columnar cactus evolution around the border of Guerrero and Michoacán. I have noted reports of plants from near Ciudad Altamirano, Guerrero, on the Río Balsas and in the Río Tepalcatepec basin, both of which drain into the Infiernillo.

The *tiponchi* is probably the easiest to identify of all columnar cacti, for its branches are topped off with incongruous but attractive yellow to brownish-orange "caps" that appear to be screwed on to the branch below. Whereas cephalia and pseudocephalia in other plants clearly emerge

Figure 2.172. Mitrocereus fulviceps, *Reyes Metzontla, Puebla.*

from the plants' tissue, the tiponchi's caps seem to be incidentally threaded on. They are formed of bristly brown, red, and yellow hairs, interspersed with occasional strands of gray. In the dry season, when most of the trees of the tropical deciduous forest of the area have dropped their leaves, these distinctive plants are visible from afar, the yellowish-red crowns resembling Russian *shliapa* caps. I have noted them from nearly a mile away. During the rainy season, the caps are visible, but the rest of the plant is nearly indistinguishable from the lush green of the surrounding thornscrub. The caps, as much as 40 cm long and 20 cm wide, often fall from the ends of the branches, which appear to be inserted inside this strange cephalium. The flowers

Figure 2.173. Tiponchi (Backebergia militaris), *Cañon Infiernillo, Michoacán.*

and fruits develop within this protective muff, which also prevents them from being easily seen. The bands of colors of the cephalia may represent a history of growth. (Figs. 2.173, 2.174, and 2.175.)

Tiponchis may reach more than 15 m in height, with thirty or more branches, but this height is unusual. In the vicinity of the Presa Infiernillo, they seldom exceed 10 m in height, the branches are decidedly thinner, and the plant is

Figure 2.174. Branches of tiponchi, Cañon Infiernillo, Michoacán.

Figure 2.175. Close-up of tiponchi cephalia.

generally slimmer in habit than either *Mitrocereus* or *Isolatocereus*. The fruits ripen in July and August, but are small (little more than 2 cm in diameter) and inconspicuous, sprouting from among the exuberant growth of colorful bristles. Local residents report that the fruits are not usually eaten, perhaps a reflection of the clear superiority of *Stenocereus friçii* and *S. quevedonis* fruits that abound in the region at roughly the same time. The fruits are also so deeply embedded within the cephalium that harvesting them would be tricky.

Locals say the wood is strong enough to serve as varas, or cross-hatching, for roofs, which is a shame given the great natural beauty of this species and its highly restricted distribution. The greater danger to the plants, though, is collectors and horticulturalists who covet tiponchis as ornamentals for their bizarre "caps." The species has been overcollected, and the tops of the branches are reportedly often lopped off to gain access to the cap, which is thought to be amenable to planting as a cutting. In the past, truckloads of the upper branches have made their way to Guadalajara and Mexico City for eager horticulturalists. The plants, though not rare within their restricted habitat, are nowhere common, and the Mexican government considers the species to be endangered—-that is, in danger of extinction. This classification may be an exaggeration, though, for the plants propagate well. In several cases, I found numerous seedlings and juveniles flourishing in the shade of the parent tree. Fallen branches may also sprout roots and become newly established. Recruitment of tiponchis requires further investigation.[116]

The Pasacana and Its Andean Associates: *Trichocereus, Stetsonia,* and *Oreocereus*

Although the columnar cacti of North America seem to lend themselves to a somewhat natural grouping, such a convenient organization is less obvious in the case of South American columnars. I have chosen to begin at the southern limits of great cacti, focusing on three genera well represented in their habitats in Argentina, Bolivia, and Paraguay.

Trichocereus atacamensis var. *pasacana*
cardón, pasacana, *kehuayllu* (Quechua), cavul

These large, thick columnar cacti are probably the most used cacti in South America. Indeed, it is doubtful that high-elevation pre-Inca cultures of the southern Andes, especially that referred to as the Dieguito, could have developed to the extent they did without them. Pasacanas are common in northwest Argentina, in adjacent northern Chile, where they are called *cavul,* and in the highlands of southwest Bolivia, where Quechua speakers refer to them as *kehuayllu.* In the Argentine highlands above 2000 m in Jujuy and Salta provinces, *T. atacamensis* is the only common large columnar cactus, and it is abundant. *T. tarijensis* crops up occasionally in the northern portions of the Quebrada de Humahuaca, but these single-stalked columnars tend to be isolated. *Oreocereus celsianus* can be found in southwest Bolivia up to 3500 m and reach a height of 6 m. In the high, dry puna in Argentina east and south of the Bolivian border, however, it rarely reaches a height that would qualify it for inclusion among the columnar elite. Pasacanas are the dominant plant at the margins of the Argentine puna. (Fig. 2.176; see also fig. 1.11.)

Pasacanas, as both the cactus and the fruits (sometimes just the fruits) are commonly known, grow slowly. Ultimately they may produce a host of arms and reach 8 m in height and 80 cm in diameter in a life span that may stretch more than three hundred years. Although the pasacanas' range is substantial, the plants have narrow ecological requirements. They cannot tolerate high humidity, extreme heat, or enduring cold. They appear to be intolerant of temperatures below -9°C (13°F) or in excess of 30°C (86°F). They demand high altitudes, with the preferred range between just above 2000 m up to just below 4000 m, under favorable exposure to the sun and climatic conditions.

On steep slopes above 3800 m at the edge of the Salar de Uyuni in Bolivia, pasacanas grow in large numbers. At this upper limit of the range, they grow only on northeast-facing slopes, where they intercept the sun's first rays each morning. At such high elevations, once the slopes merge into a plateau, the cacti disappear. The more direct angle of the morning sun on the slopes invites more rapid warming of the surrounding substrate and air than is possible on the flats. Similarly, the slopes receive the benefit of convection of warm air from the lower valleys, a benefit absent on the plateau surfaces. Increasing numbers of individuals at the highest elevations tend to remain a single stalk rather than producing branches, especially in Salta and Jujuy, Argentina, and mature individuals are usually much shorter than plants in lower and more benign habitats. Similarly, many individuals at the highest elevations sport dense coats of white fluff, far more so than at lower altitudes, perhaps as protection from frost and sunburn. At Quilmes, Argentina, at around 1900 m on a gentle slope, a majority or near majority of the mature plants are multibranched, whereas at Las Pailas, on a rocky alluvial fan at 2900 m, only a small proportion of the mature plants has branches. Near Cuesta

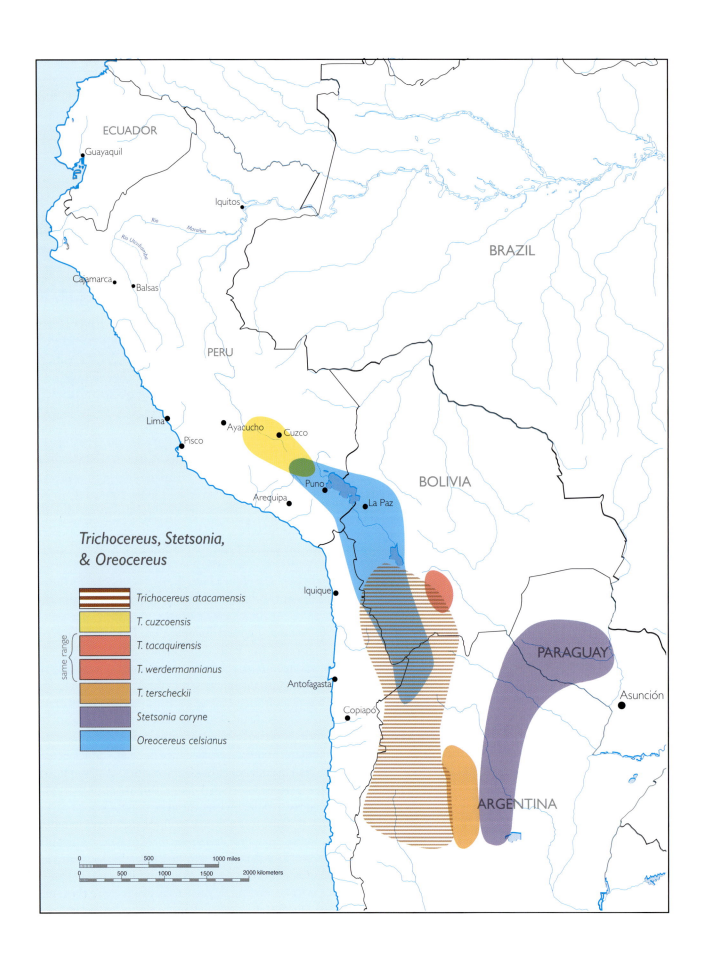

Trichocereus, Stetsonia, & Oreocereus

Trichocereus atacamensis
T. cuzcoensis
T. tacaquirensis } same range
T. werdermannianus
T. terscheckii
Stetsonia coryne
Oreocereus celsianus

Figure 2.176. Pasacanas (Trichocereus atacamensis) at Quilmes archaelogical site, Tucumán, Argentina. Large plants like these are often associated with pre-Columbian ruins.

Figure 2.177. Pasacana, Quebrada del Toro, Salta, Argentina.

de Lipán at 3500 m, branching is more common, but the plants appear dwarfed, not exceeding 3 m in height. At the edge of Bolivia's Salar de Uyuni at 3760 m, many plants are branched, but are similarly dwarfed. Their upper limit at roughly 20°S latitude, however, is considerably higher than in Salta Province, Argentina, at 25°S. (Fig. 2.177.)

Pasacanas respond well to irrigation, growing up to four times as fast as they do without it, according to residents of the region. Abundance of rainfall per se, then, is not a limiting factor in their distribution. Still, the plants are not found in areas with more than about 200 mm of annual rainfall. This detail suggests that humidity and perhaps soil moisture are the most important limiting factors, with relative heat and cold playing a secondary role. Explosions of native pests can also decimate populations. On the *pucara* (archaeological site atop a hill or mesa) of Tilcara, Jujuy, Argentina, an extensive population of pasacanas reveals signs of stress

(especially those individuals on the southern exposure of the fortification). Recent studies of plants on the Tilcara pucara suggest that this nearly universal pathology is likely due to rapid proliferation of a coddling moth whose populations previously were kept in check by a predatory wasp. The wide use of pesticides in adjacent fields may be responsible for a sudden drop in predator populations, leading to an explosion of coddling moths that threaten the survival of nearly all the thousands of plants on this and other pucaras of the Quebrada de Humahuaca.[117] Elsewhere, plants free of contact with pesticides appear to be flourishing.

High humidity appears to limit the distribution of pasacanas. In approaches from the east to Los Cardones National Park in central Salta province, the first plants are found only where the prevailing fogs dissipate, at 2800 m some 20 km west of the rim of the plateau, and those first individuals appear stunted and stressed. As rainfall, fog

Figure 2.178. Pasacanas, Quebrada del Toro, Salta, Argentina.

cover, and elevation decrease to the west, pasacana numbers increase, and the cacti's stature and vigor are enhanced. In contrast, perhaps the most vigorous large population occurs at 2900 m at Las Pailas ruins in the upper Cachi Valley, where precipitation is higher than elsewhere in that region, but fog is relatively uncommon. Bolivian pasacanas begin to disappear, even at high elevation, as one travels eastward into areas of higher rainfall and corresponding greater humidities. But pasacanas also appear to require at least 75 mm, or 3 in., of rainfall, roughly the same as that of Yuma, Arizona, which is generally too dry and perhaps too hot for saguaros. Pasacanas raised in the Sonoran Desert environment of southern Arizona require considerable shade in order to survive the heat. Temperatures there are 10–15°C (50–59°F) higher than any occurring in the plants' native environment.

In spite of these narrow ecological requirements, pasacanas are very common in the Argentine and Chilean Andes.

Indeed, at higher elevations, they are often the only native plant noticeable from a distance in the landscape, and their numbers appear limitless. They cascade down steep hills, spilling onto alluvial fans like divisions of soldiers. On vast mountainsides and tops, they grow widely spaced but ubiquitous. They prosper among boulders on coarse granitic soils and flourish on bajadas in alluvial sediments. (Fig. 2.178.)

Pasacanas have white flowers that grow only on the north side, where studies have noted daytime temperatures as much as 18°C higher than on the cooler south side.[118] The flowers open at night and are probably pollinated by bats. At lower elevations, juvenile plants (those that have yet to flower) have very long and formidable russet spines (in excess of 20 cm) that lend the young cacti a distinctly reddish hue, especially near their crowns. In addition to offering protection against the predations of herbivores, these remarkably long, potent spines may provide the youthful

Figure 2.179. Spines of Trichocereus atacamensis, *Los Cardones National Park, Salta, Argentina.*

plants with an increased surface area for condensation of dew. Surface tension then retains the dew on the surface of the spines. The dew drains toward the ribs of the plant and subsequently downward to the base, where it provides increased moisture to the roots. Mature individuals, in contrast, have white spines at the crown, usually no longer than 5 cm long, and the upper portion of the plants typically appears white from a distance, providing the great cacti with a distinct corona when backlit. (Fig. 2.179.)

The pulp of the pasacanas (the fruits) is white, and the black seeds are similar in size, though slightly smaller, to pitaya seeds *(Stenocereus thurberi)*. The pasacanas ripen in March and may remain attached to the plant long after they have dried, well into May in southern Bolivia. A plant's age can be roughly determined by counting ersatz growth rings—the areoles on each rib.[119] Fruit scars are parallel to each other all the way around the cactus, and one such ring is produced each year. The external tissue between the growth scars also records a history of rainfall in the vertical lines of the ribs. Indentations or concavities indicate dry spells, and convexities indicate times of greater moisture.

The plants recruit well when they are free of the depredations of livestock, especially burros and cattle, which may eat or trample seedlings not completely protected. At higher elevations in Bolivia, there is insufficient forage for cattle, and temperatures are rather cold. Recruitment of kehuayllas is excellent there. Foraging llamas appear to avoid the plants, and their more delicate hooves do not appear to compact adjacent soils. The fruits are virtually the only wild fruit available and are much relished by native peoples.

A dense grove or group of pasacanas clustered in an area is often an indicator of archeological sites, especially pucaras or fortifications. Exceptions seem to be the extensive ruins of Los Amarillos, west of San José, Jujuy, in the Quebrada de Humahuaca, where a mountainside site of tilted strata reveals only scattered pasacanas and the vast ruins at Tastil at 3100 m, where the lack of soil and dry conditions do not appear to be conducive to *any* plant growth. The soil in both places is thin and quickly eroded, making for poor pasacana habitat. Elsewhere, the seeds of the fruits in these once heavily inhabited areas passed through the human digestive tract, thus promoting germination. (The aborigines did not possess domestic animals that would discover and eat the seeds or recently germinated young plants, so the possibility of dispersion by large mammals can be rejected.) The feces were concentrated in areas of intense human occupation. The piling of stones and large rocks that accompanied the construction of human fortifications and dwellings provided excellent habitat for the cacti. They flourished in the coarse, rocky, well-drained soils while benefiting from the mass of rocks that acted as a heat sink, protecting them from temperatures that may reach -9°C (18°F).

Without cactus fruit in their diet, aboriginal peoples of the desert and thornscrub of northwest Mexico could not have survived. I suspect that the same is the case with the pasacana for peoples of the puna. The fruits may have constituted the principle source of natural sugars in the local diet prior to the advent of prolific honey producers (honeybees) and sugarcane. If so, the presence of many pasacanas nearby would be a favorable development, supplying both the sweet fruit and lumber. Steep slopes around the Salar de Uyuni in southwest Bolivia harbor fine stands of the plants. The slopes are also home to numerous pre-Columbian *chullpas* (burial caves), recently pillaged, many of which incorporate beams of pasacana wood to maintain open passageways. The beams show no sign of decay. The Incas and their predecessors were sufficiently confident of the pasacana wood's strength and durability that they entrusted it to maintain their sacred sites for eternity.

From a distance, healthy lower-elevation pasacanas superficially resemble the saguaro of the Sonoran Desert of Arizona and Sonora, which also has more than twenty ribs, whereas the lower-elevation *T. terscheckii,* with fourteen to sixteen ribs, more resembles the sahueso of Baja California and the Gulf of California coast of Sonora. At times, individual pasacanas will develop in excess of twenty arms. These multibranched and sometimes multitrunked individuals often grow near roadways (reflecting the highway shoulders' ancient role as toilet!). The proliferation of branches may be a response to insults to the plant by passersby. In these multibranched individuals, it is not uncommon for the branches to grow densely vertical, often parallel to each other with no space between them, a habit almost unknown in saguaros and sahuesos, but rather common in tetechos (*Neobuxbaumia tetetzo*) of southern Mexico.

The pasacana fruits, ripening in March, are gathered (or were so until recently) and eaten with enthusiasm. Rafael Liquitay of Tilcara recalled that when he was young, people would say, "Vamos a pasacanear," and off they would go to collect the fruits. The fruits are sweet and satisfying, he said. People reportedly still gather fruits in the puna region of western Salta province.[120] Rocha (Rosario Corimayo) of La Paya, Cachi, recalled someone long ago who would bring in a batch of pasacanas, collect and dry the pulp, and make *queso de pasacana,* something akin to guava paste, using the same native grass *cinchón* (*Lamprothesis* sp.) mold that was used for forming wheels of cheese. Rocha also believes his ancestors may have made wine from the fruit. He reported that the husks from the fruits were formerly burned and the ashes mixed with potatoes to make a *yista* for catalyzing coca leaves, which he and his ancestors have chewed since time immemorial. As of the late 1990s, yista was still prepared from pasacana husks and flowers and sold commercially in local markets in Jujuy, Argentina. (Coca leaves are legal merchandise in the provinces of Jujuy and Salta.) Residents of the Bolivian Altiplano community of Santiago K at 3900 m, southwest of Salar de Uyuni, still feast on the fruits, even though they must travel considerable distances (6–10 km) to gather them. Rather than accumulate piles of the fruits, they customarily eat the fruits first gathered on the spot, scooping out the sweet mass with a spoon. Only after eating several do gatherers collect additional fruits for taking home.

The rot- and insect-resistant wood, though porous and marked with holes where the areoles intrude, is surprisingly hard and durable. It finds its way into many highland homes, especially in ceilings, roofing, flooring, furniture, and other carpentry, and in artifacts such as chests, boxes,

Figure 2.180a. Chair made of pasacana wood, Tupiza, Bolivia.

and picture frames. Local producers offer logs and planks for sale along highways in Salta and Jujuy. In southwestern Bolivia, nearly every home sports an article made from the pasacana wood, which is often the only straight-grained lumber available. Furniture made from the wood is sold in markets and shops throughout the southwest Andes. (Figs. 2.180a and 2.180b.)

Although sawing cactus wood into planks and boards seems somehow improbable or odd, producing lumber from the cactus is mechanically no different from sawing up other hardwoods, except that the trunks are hollow, with a core from 10 to 30 cm in diameter. The wood surrounding this hollow core is up to 10 cm thick and can thus be ripped into planks 5 cm thick by 25 cm wide by 2 m long. The trunk can also be sawed crosswise into posts, like other ordinary trunks, yielding posts up to 2 m tall. The wood is tough enough even to have been used as flooring and has proved as durable as the best hardwoods. I was skeptical about the wood's qualities,

Figure 2.180b. Interior of church, Purmamarca, Jujuy, Argentina. The pews, balustrade, and ceiling panels are made of pasacana wood.

but was converted to a promoter by a sturdy floor more than one hundred years old in a historic building in Cachi, Argentina. In several churches of the Salta and Jujuy highlands, the predominant lumber used throughout the buildings (except for beams, which are of mesquite) is pasacana wood, including the pews, which are surprisingly comfortable. Lecterns and pulpits in churches often feature a base, column, and stand of pasacana wood. The confessional of the Cachi chapel is constructed of a softwood (perhaps pine), with a veneer of the harder and more durable pasacana. Ceiling panels in many buildings are of thin, wide strips of pasacana nailed between beams of native mesquite or *quebracho* (*Schinopsis hankeana*, Anacardiaceae) from the lowlands. In Tupiza, Bolivia, at 3000 m, somewhat below the range of pasacanas in that valley, household furnishing such as chairs and tables are commonly constructed from pasacana wood, which is praised for its durability and workability.

Figure 2.181.
Fence of
pasacanas
protecting
irrigation ditch,
Cachi, Salta,
Argentina.

No columnar cactus of northwestern Mexico or the southwestern United States produces wood that possesses the characteristics of such fine lumber. Etcho *(Pachycereus pecten-aboriginum)* lumber is used in construction and for making plank doors and wooden beds, but it does not receive the widespread incorporation that pasacana does as a lumber source in Argentina, nor is it routinely sawed into boards for construction. However, pasacana wood was the only lumber available for its uses, except for several mesquite species, the wood of which possesses an irregular grain and is hard and difficult to work; the trees also often grew at a considerable distance from pre-Columbian construction sites.

To produce such volumes of lumber, local woodcutters centuries ago learned to girdle the cacti and wait for them to die—perhaps for several years. When the individual *T. atacamensis* tree dies, it remains upright. Gradually the outer skin rots and falls away, according to Rafael Liquitay of Tilcara. When the remaining bark is peeled off, the underlying wood is still resilient and durable. The volume of wood harvested makes me somewhat nervous for the pasacana's future—after all, it is a very slow-growing plant—but local producers assure me the supply is ample.

Rocha also told me that the long spines of the juvenile plants were formerly used as needles to darn socks and stockings. I have snipped off spines more than 20 cm long from plants and will testify that they are nearly as tough as steel needles. Agriculturalists and pastoralists of Bolivia's high altiplano still use the spines for sewing and shape them into strong, effective, and free fasteners as well.

Well into the twentieth century, natives of the Argentine puna planted living fences of pasacanas that still perform their function. Rocha pointed out that cuttings readily sprout when dried for two or three days and then planted 30 cm deep in moist soil. If they are planted at the onset of the rainy season, they will sprout in that same season, he said. If the tip of a living plant is sliced off and planted, it will continue to grow as a single stalk. If the cutting is a section of a trunk, it will immediately branch upon being planted. As proof of the cutting technology, Rocha pointed out a corral-like rectangular fence his grandfather had built in roughly 1940. The handsome fence encloses a small field and garden to protect them from livestock. His grandfather, who lived to be 110, supplied the cuttings with irrigation water from catchments he constructed nearby. The tallest of the plants is now roughly 6 m tall, and many have sprouted arms, indicating a rate of growth faster than that of saguaros.[121] In the Valle de Cachi Adentro, numerous similar cardón pasacana fences still flourish that were long ago planted adjacent to the road and associated with *acequias* (irrigation ditches). (Fig. 2.181.)

In times of severe drought, livestock growers may burn pasacanas to scorch off the spines, fell them, cut them into sections with an axe, and leave the fallen giants for livestock, who find the spineless trunks marginally palatable.

Although this persecution and the girdling of many plants for lumber paint a dire picture, recruitment of juveniles appears to be widespread, and a healthy population seems to assure the plant's continuance for many centuries, as long as diseases such as that experienced in Tilcara do not proliferate.

Some cactologists have recently referred *Trichocereus* to *Echinopsis.* This move influences the taxonomy of South American cacti because the tribe Trichocereeae is one of three basic divisions in the continental phylogeny. The whole apparatus would have to be replaced or redefined.

Trichocereus cuzcoensis Britton & Rose
San Pedro macho, sanquey

Botanists in the Ayacucho region of Peru label this columnar cactus *T. peruvianus.* Its description is somewhat at odds with that of Anderson, however, so I identify it only tentatively. It is probably the most widespread and common columnar cactus in the Ayacucho region, thriving between 2000 and 3500 m elevation on the dry slopes.

The plants are nondescript but quite spiny, reaching a height of at least 8 m, with many branches. They grow abundantly on the vast ruined site of Wari, the location of a great city that flourished near Ayacucho between 700 and 1100 A.D. They can also be found growing in fields at Pomaqucha, the artificial lake created by the Incas high on a mountainside, suggesting that the Incas as well as the Waris planted them. They are also planted in yards and in pots throughout the area. The plants are common on hillsides throughout the region, as well as in yards, where they produce fruits that are widely eaten. A shampoo is reportedly made from the flesh of the cactus, but the methodology remains murky. Rumors also persist that the branches contain mescaline in sufficient concentration to induce hallucinations when properly administered. (Figs. 2.182, 2.183a, and 2.183b.)

Trichocereus tacaquirensis (Vaupel)

This small columnar cactus grows in the high valleys of southern and central Bolivia. I know nothing more about it. It reaches more than 5 m in height. (Fig. 2.184; see also fig. 2.192.)

Trichocereus terscheckii (Parmentier) H. Friedrich & G. D. Rowley
cardón, South American saguaro

Larger than the pasacana is its parallel companion *Trichocereus terscheckii,* parallel because the latter occupies habitats found at elevations immediately below those at which

Figure 2.182. Sanquey (Trichocereus cuzcoensis) *growing on ruins of the ancient city of Wari, Ayacucho, Peru. Elevation around 3000 m (10,000 ft.).*

the pasacana grows. *T. atacamensis,* at the lower elevational limit of its distribution (around 1500 m), appears to hybridize readily with the larger *T. terscheckii.* The latter cardón (I have been unable to determine an indigenous name for the plant) has entirely different ecological requirements from those of the pasacana, so much so that their intermingling is usually brief and over a narrow band. Still, *T. terscheckii* seems quite promiscuous, and the areas where intergrading occurs demonstrate numerous phenotypes of hybrids living harmoniously with nonhybrids. Pasacanas are far less common in some sites of hybridization, suggesting that *T. terscheckii* is truly a wanton species, as if it were struggling to gain the genetic wherewithal to grow at higher elevations. (Fig. 2.185.)

Hybridization occurs among other members of the genus *Trichocereus,* the hybrids producing fertile offspring.[122] In the Argentine province of Mendoza, human activities such as grazing and trampling by domesticated livestock and the gathering of firewood have favored the proliferation of cacti, which may well have affected the proliferation of

Figure 2.183a. Sanquey in cornfield near Inca resort of Pomaqucha, Ayacucho, Peru.

Figure 2.183b. Trichocereus cuzcoensis *in field of* Opuntia *sp., Wari, Peru.*

Figure 2.184. An unusually tall Trichocereus tacaquirensis, *Limeta Valley, southern Bolivia.*

Figure 2.185. Trichocereus atacamensis *X* terscheckii *hybrid, Quebrada del Toro, Salta, Argentina.*

hybrids as well as the dense populations of both species. The extent of fertility of the hybrid *T. atacamensis* X *T. terscheckii* has yet to be studied.

T. terscheckii appears to have struck a deal with pasacanas, in which each has its own ecological niche and does not intrude in the other's domain. The upward-elevation range of *T. terscheckii* is limited by cold and decreasing rainfall, whereas the lower range of the pasacana is limited by heat and increasing rainfall. Specimens of *T. terscheckii* do not appear above roughly 2000 m, although below this point they vastly outnumber specimens of *T. atacamensis,* which rapidly drops out of the picture. In the profound Quebrada del Toro of Salta province, Argentina, they flourish on steep hillsides and bajadas, sometimes growing in thick stands, with numerous plants approaching 10 m in height and often serving as hosts to ballmosses (*Tillandsia* spp.). Plants that have apparently hybridized with *T. atacamensis* tolerate slopes of lower elevation better than the pure strain of *T. terscheckii* does. Above 2000 m here, as elsewhere, however,

freezing is more common and appears to limit the plants' upward mobility.

Cardones (as I call *T. terscheckii*) are often massive plants, up to 10 m tall and usually with numerous large arms. They seem to prefer the semiarid intermediate Chaco forest habitats that receive 400–750 mm annual rainfall, but their densest populations are on the upper periphery of that plant community (perhaps due to human-related removal of competing species). They can also be found in verdant semitropical deciduous forest, their tops exceeding the surrounding vegetation, an accomplishment apparently essential for their survival. Their lower-elevation limit seems to involve their ability to withstand competition from the Chaco forest and from the taller trees in the *yungas,* the cloud and rain forests that cover the eastern slopes of the steep foothills of the Andes. They have an affinity for steep hillsides, are seldom found growing on flatlands, and are less common on gentle slopes. (Fig. 2.186a.)

Figure 2.186a. Trichocereus terscheckii, *Salta, Argentina.*

Figure 2.186b. Trichocereus terscheckii *growing from* algarroba *(*Prosopis *sp.), Salta, Argentina.*

The plants flower at various times of the year, but primarily during the rainy season, December–March. The large white flowers (14 cm) open at night and appear to be pollinated by bats.

These cardones range well south into the Argentine provinces north of Mendoza, including Catamarca, La Rioja, San Juan, and Tucumán, as far south as saguaros range north, always growing at elevations below the pasacanas. Unlike pasacanas, however, they do not cross the Andes. They are found only in Argentina.

As prominent as these huge cacti are, I was unable to discern many human uses for them. Residents of their habitat have reported that the fruits are edible, but inferior to those of the pasacana. They appear to be important fodder for livestock, especially goats, functioning as a time-release food as they drop to the ground over an extended period of time. The wood appears to rot rather quickly and lacks the characteristics of good lumber. Still, local people are said to fashion rustic furniture from the branches, but I have

been unsuccessful in locating any examples. I speculate that because *T. terscheckii* shares the habitat of scores of other trees and shrubs (unlike *T. atacamensis,* which is often the only large plant and the only fruit source in its habitat), it lacks any particular characteristic that cannot be better met by other plants in its community.

T. terscheckii is a fast-growing plant, especially when compared with *T. atacamensis.* That three plants managed to become established and thrive in a mesquite tree near Salta, Argentina, indicates their hardiness. One *T. terscheckii* planted in my yard in Tucson grew from 20 cm to nearly 2.5 m tall and sprouted two branches over eight years. The plants have gained considerable popularity in the United States, where they are known by the name "South American saguaro." They tolerate frost quite well and require only minor supplementary watering. Horticulturalist Dan Bach of Tucson reports that a customer successfully raised one outdoors in Atlanta, Georgia, a place not widely known as a haven for cacti. (Fig. 2.186b.)

Trichocereus werdermannianus Backeberg
cardón

I know very little about this large, handsome plant except that it is common in a stretch of the Limeta Valley in southern Bolivia, where it is well represented on hillsides at about 3000 m elevation. This cactus reaches 8 m in height with a thick trunk and resembles *T. atacamensis* in habit, but is more massive and has comparatively few spines. Its fruits should be tasty. (Fig. 2.187.)

Stetsonia coryne (Salm-Dyck) Britton et Rose
unquillo, cactus, cardón, tuna, toothpick cactus

An additional columnar cactus from Argentina is noteworthy. *Stetsonia coryne* is a many-branched columnar cactus of the lower valleys (700 m and below), the Chaco forest of the northwest lowlands, and similar habitats in Bolivia and Paraguay. Among nonindigenous people, it is usually referred to simply as *cardón* or *tuna*. In eastern Salta, it is also referred to as *unquillo,* but the general applicability of that name is questionable.[123] I use this name simply to avoid the monotonous name *cardón,* which applies to at least a dozen species of columnar cacti. (Fig. 2.188.)

In parts of the Chaco forest, the unquillo forms the same dense groves characteristic of the two giant *Trichocerei.* Throughout its range, it is common to very common. In outline, the plants resemble somewhat the many-branched *Polaskia chichipe* and *P. chende* of Mexico's Valle de Tehuacán, branching quickly and prolifically from a short trunk to present an overall appearance of an inverted triangle set on a post. The no-nonsense spines are white, straight, sturdy, and sharp—up to 10 cm long. They are the basis for the common English name *toothpick cactus.*

Unquillos bloom in January and February, and they fruit in March and April. The large white flowers open at night and are thus probably pollinated by bats. Near the city of General Güemes, Salta, the plants reach a height in excess of 7 m. In the Chaco, they are said to reach even greater heights, up to 9 m. Oddly enough for a large cactus, plants sprout readily from cleared land, and colonies form part of the first succession of the plant community. Thus, they also appear to benefit from heavy grazing, for once they are established and send out their potent, long spines, cattle leave them alone. The unquillo demands relatively flat lands, tolerating a wide range of soils, including dense clays that deter other shrubs and trees. It is almost totally absent from slopes. It also seems to flourish in saline soils, though the plants growing in salty soils sometimes look scrofulous. Its ability to sprout readily on severely overgrazed pastures makes its future in the heavily grazed Chaco

Figure 2.187. Trichocereus werdermannianus, *Limeta Valley, Bolivia.*

bright indeed. However, buffelgrass *(Pennisetum ciliare),* an aggressive African forage grass, has been introduced into the region in recent years, and its penchant and need for fire may pose a dire threat to the unquillos' continued prosperity. (Fig. 2.189.)

Unquillos form an important contribution to human cultures, especially in the Gran Chaco. The fruits, appearing in late summer, are smooth skinned and have a lemony flavor. They appear to be one of only a few columnar cacti fruits whose rind is edible raw. The mass of black seeds embedded within white pulp is usually cooked, another unusual characteristic of the fruits, but can also be eaten raw. In addition to the fine and highly edible fruits (that form an important forage source for livestock), the wood is used to make furniture. The inner cortex dies and rots when the plant dies, leaving an attactive, often straight cylinder, ideal for the "rain sticks" popular with tourists.

The taxonomic status of *Stetsonia* is uncertain. Wallace suggests that its traditional inclusion in the tribe

*Figure 2.188.
Toothpick cactus*
(Stetsonia coryne)
*near General
Güemes, Salta,
Argentina.*

Figure 2.189. Stetsonia
coryne *plants in Chaco
vegetation, Salta,
Argentina.*

Figure 2.190. Fruit of Oreocereus celsianus, *Potosí, Bolivia.*

Figure 2.191. Oreocereus celsianus *near Tupiza, Bolivia, elevation around 3600 m (12,000 ft.). The dense coat of hairs affords protection from freezing.*

Browningieae is perhaps mistaken and that its affinities are more likely with the tribe Cereeae, the cacti of South America east of the Andes, where it grows.[124] At any rate, it is only a distant relative of the two columnar *Trichocereus (Echinopsis)* species.

Oreocereus celsianus (Lemaire) Riccobono
cardón; old man of the Andes

Once again I am embarrassed to present a handsome columnar cactus known only by the bland name *cardón*. This species is noteworthy for the dense mat of hair that covers the branches, so thick that it conceals the very effective spines. Its flowers are an attractive pink color shading

to grape, and its fruits are lime green with a roundish, sometimes dimpled rind. The fruits are unusual in that they appear to have little in the way of pulp. When they are cut open, their insides are reminiscent of a green pepper. The seeds are edible, but few. (Fig. 2.190.)

These cardones are high-altitude plants, the large specimens found in southern Peru, Bolivia, and Argentina above 2500 m, reaching at least 4000 m, higher even than pasacanas. They are often quite common at these heights, and in the pure air of the high Andes at certain hours of the day they seem to illuminate the landscape with the many white tips of their branches that resemble points of light. They reach nearly 6 m in height, with sturdy, thick

Figure 2.192. Oreocereus celsianus *(center) and* Trichocereus tacaquirensis, *Limeta Valley, Bolivia.*

branches and trunks. They are probably the hairiest of thick-trunked columnar cacti. The hair serves an important insulating function, providing a mat of interwoven fibers that protects the growing tip and other parts of the cactus from the frequent freezes that occur at high altitudes throughout its range. (Figs. 2.191 and 2.192.)

Espostoas

Espostoa is named in honor of Peruvian Nicolás Esposto, a botanist with the National School of Agriculture in Lima. This remarkable Andean genus ranges from southern Ecuador south to Bolivia, with the bulk of the species centered in northern Peru. Plants are known only west of the Andes, with a concentration of species along the Rio Marañon, the odd river that flows northwesterly through a profound canyon west of the massive Cordillera Oriental before breaching through a spur of the Andes and flowing east into the Amazon. The genus *Espostoa* contains between twelve and sixteen species, depending on the authority one

wishes to cite. At least six species make their home in the region traversed by the Río Marañon and its tributaries, and because vast areas bordering the river are inaccessible, the number of species is subject to change: Charles and Woodgyer described a new species in 2003.[125] The taxonomy of the genus is still in flux, so the identifications given here are tentative. I have relied heavily on Charles's work for the names I have assigned to the plants. In spite of the isolation of the home territories of most *Espostoas,* their attractive hairy branches have made them horticulturally popular throughout the world. Jens Masden introduced me to the genus and was kind enough to accompany me into the field in Ecuador and point out the two species that grow in that country.

Most *Espostoas* have white flowers[126] (*E. lanata*'s are pink-purple) and open at night. They lie embedded within the hairy cephalia (or pseudocephalia). I have noted some flowers on plants growing in my yard that seem to remain open during the day. I am accustomed to admiring the large, wide-open flowers of columnar cacti and the succulent

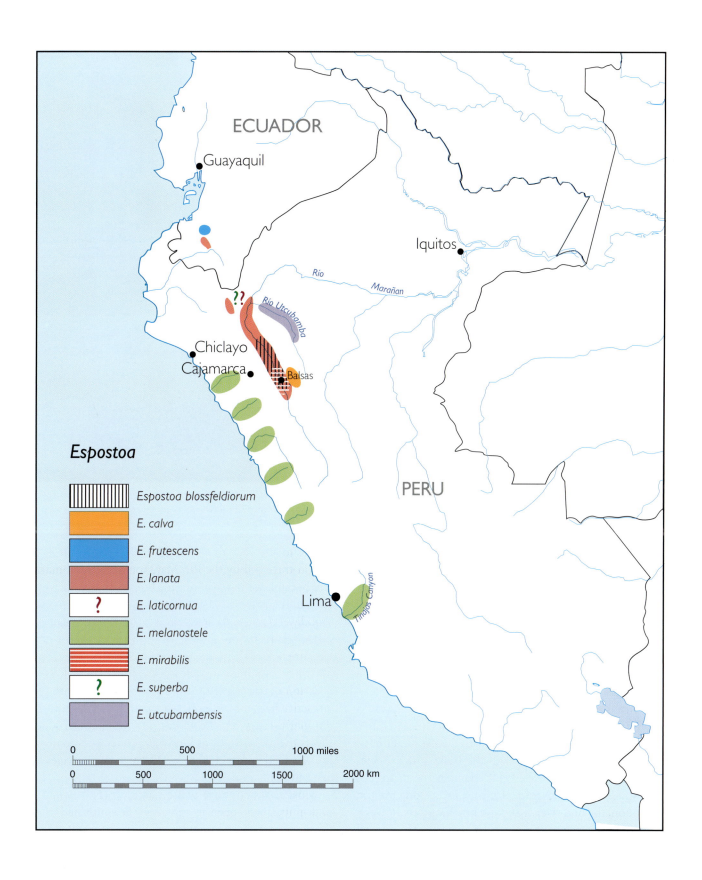

Espostoa

⦀	*Espostoa blossfeldiorum*
🟧	E. calva
🟦	E. frutescens
🟥	E. lanata
?	E. laticornua
🟩	E. melanostele
≡	E. mirabilis
?	E. superba
🟪	E. utcubambensis

ECUADOR

Guayaquil

Iquitos

Río

Marañan

Río Utcubamba

Chiclayo

Cajamarca

Balsas

PERU

Lima

Tingjas Canyon

0   500   1000 miles

0   500   1000   1500   2000 km

fruits. In comparison, both flowers and fruits of *Espostoas* seem small, but the plants and species have proliferated throughout the Marañon region and its environs, so they must be carrying out their task of reproducing rather well. So far as I can tell, *Espostoa* is an aberrant genus in that it does not produce tasty or sweet fruits. The few fruits I have observed are edible, but of odd texture and a flavor that ranges from indifferent to strange. (Fig. 2.193.)

Most *Espostoa* species exhibit varying densities of lateral cephalia; that is, beds of woolly hairs grow on the sides of the branches, often making them resemble elongated leeches. In some species (*E. guentheri* of Bolivia), the lateral cephalia expand to encircle the branches. Most of the species grow to be handsome adults, some downright beautiful. The cephalium is often oriented in only one direction, but this characteristic appears to me to be inconsistent among plants, and even in individual plants different branches will have cephalia facing in different directions. Given *Espostoa* habitats' proximity to the equator, there is no obvious advantage to orientation in any particular direction.

Other than the few uses of the plants given in the species descriptions here, the genus does not have a varied and wide ethnobotanical report card. One possible reason for this (apart from the reason that the genus may not be useful!) is the absence of indigenous people throughout the bulk of its range. I have found that ethnobotanical uses of native plants tend to decrease sharply in the absence of peoples with many centuries of land tenure. Pre-Columbian peoples were largely exterminated or assimilated by Spanish colonists in the Marañon region, northern Peru, and southwestern Ecuador, and their descendents and the indigenous peoples identifiable as such ceased to occupy the *Espostoa* habitats centuries ago, a fact that may help explain the absence of human uses of this fine genus.

Ecuador is home to at least two species, *E. frutescens* and *E. lanata,* the latter growing in Peru as well. Both are notable for the whitish tint produced by the dense woolly hairs on the upper arms. *E. guentheri* is found in central Bolivia. I am not aware of having seen it. The remaining species are Peruvian, the majority found in northern Peru.

Espostoa frutescens J. E. Madsen

E. melanostele (Vaupel) Borg.

E. lanata (Kunth) Britton & Rose

Espostoa frutescens is endemic to a couple of dry, narrow valleys in southern Ecuador, probably limited to a narrow belt of desert in each. It is smaller than *E. lanata,* but equally pilose, its woolly heads clearly visible against the dry desert background. Although it barely clears 2 m in

Figure 2.193. Flower and fruit of Espostoa lanata, *Catamayo Valley, Ecuador.*

height, I classify it as a columnar. *E. frutescens* is a most handsome plant. It appears quite abruptly along the highway from Cuenca to Pasaje. West of Santa Isabel the route drops precipitously from the central highlands toward the lowlands and a moist montane forest climate gives way in a few kilometers to a dry desert. As the visitor traverses the canyon, *E. frutescens* and *Pilosocereus tweedyanus* appear suddenly and in impressive numbers, evident on all sides for a brief stretch, then vanish from the scene as abruptly as they appeared. The canyon sides close in and quickly become green once again, and the desert vegetation merges into cloud forest. *E. frutescens* may have evolved in this discrete, sudden band of aridity, a most satisfying brown desert Eden (for a desert lover) surrounded by saturated verdure. (See fig. 1.47.)

I was unable to obtain a common name or information about uses of *E. frutescens.* No one appears to live in the region where it is found, understandable because although

Figure 2.194. Espostoa melanostele, *Cañon Tinajas, Peru.*

upcanyon and downcanyon environments are entirely hospitable and well watered, the brown, desiccated slopes of the *Espostoa* desert appear singularly hostile to human occupation.

E. melanostele may be the most widespread of *Espostoas.* Peruvian cactologist Dr. Carlos Ostolaza of Lima took me to a desert canyon called Tinajas southeast of Lima where many of these handsome plants grow at about 1500 m elevation. He pointed out that they grow at least as far south (160 km) as Pisco and north into the Río Jequetepeque valley. Nearby are healthy *Browningia candelaris* and *Armatocereus matucanensis* plants. Far down in the valley, where vegetation is almost nonexistent and it is too dry for *Espostoas,* large, healthy *Neoraimondia arequipensis* plants appear occasionally on hillsides. (Fig. 2.194.)

E. melanostele plants in the vicinity of Lima are no taller than the Ecuadorian *E. frutescens,* yet they are strikingly handsome and surely a prime candidate for horticultural use. Around the Río Jequetepeque, more than 700 km to the north, great numbers of these plants reaching 3 m in height (but usually less) flourish on dry granitic slopes. They are an attractive cactus, the hoary hairiness providing them a sturdy and more or less clean-cut appearance, especially in contrast with the often grubby appearance of nearby *Armatocereus* and *Haageocereus,* with which they are frequently sympatric, and the large but at times sickly-looking *Neoraimondia* that grow sporadically on the steeper slopes. The fruits are tiny—hardly bigger than a grape—and piquant, but tasty and edible. Although I have read that the fruits are consumed locally, none of the natives with whom I have spoken appear to harvest them. (Fig. 2.195.)

E. lanata is a many-armed cactus, the arms rising from a common base, sometimes with a trunk. Madsen reports specimens reaching 7 m in height,[127] but those growing in the Catamayo Valley of Ecuador are little more than

Figure 2.195. Espostoa melanostele, *Río Jequetepeque valley, northern Peru.*

Figure 2.196. Espostoa lanata, *Balsas, Río Marañon, Peru, elevation around 1500 m (5000 ft.).*

half that. In the Río Marañon valley of Peru, they grow in impressive numbers, reaching at least 7 m in height and sometimes assuming a candelabra shape, dominating the landscape for many kilometers. Madsen writes that until recently the woolly hairs were scraped from the arms (which had been amputated) and used to stuff pillows. Although this practice appears to have fallen into disuse, numerous plants exhibit scars where arms have been lopped off, the limbs apparently sacrificed for human comfort. This use appears to be widespread for several species of *Espostoa*, but I could find no one to give me a demonstration of the technique. Those with whom I spoke invariably chuckled and reported that people *used* to stuff pillows with the *lana* (wool). The common name for *E. lanata* in Ecuador is *zoroco.* The fiber and the cactus are referred to as *zonca* in northern Peru. (Figs. 2.196 and 2.197.)

The distribution of *E. lanata* appears to be limited to dry valleys in southern Ecuador and northern Peru.

Espostoa blossfeldiorum (Wedermann) **Buxbaum**

E. calva F. Ritter

E. laticornua Rauh & Backeberg

E. mirabilis F. Ritter

E. superba F. Ritter

E. utcubambensis G. J. Charles

At least six (and as many as seven) additional *Espostoas* can be found in the Río Marañon region in Peru, three of them large, two of them giants. Residents of the region refer to them indifferently as zoncas, and natives report of all species that the wool was formerly harvested to stuff pillows. No one with whom I spoke mentioned eating the fruits, although some people appear to have heard *others* speak of eating them. Ethnobotanical uses of the genus are hard to come by in the region.

Figure 2.197. Espostoa lanata, *Catamayo Valley, Ecuador. Cactologist Jens Madsen is working on the plant.*

E. blossfeldiorum is a small columnar, hardly exceeding 2 m in height in single stalks or with a couple of branches joined at the base. I found it near Balsas on the east side of the Marañon and discovered it to be easily distinguished from *E. mirabilis,* which is found nearby. *E. blossfeldiorum* typically sports a bushy mound of beardlike hairs surrounding the base, a characteristic absent in *E. mirabilis.* The cactus *E. calva* is much taller, reaching a height of 8 m on hillsides well above the Río Marañon, and has yellowish spines. *E. mirabilis,* in contrast, is extremely common on the west side of the Río Marañon above Balsas, growing in the company of large *Browningia pilleifera* and *Armatocereus rauhii* and several species of smaller cacti (including *Melocactus* spp.) in great numbers and in gorgeous array. *E. mirabilis* often exceeds 4 m in height, the branches springing from ground level from a short trunk. Its ribs tend to grow in a slightly spiral form, adding an individual appearance to the plants. Its lateral cephalium is a strong russet

color that contrasts most agreeably with the dark green of the branches. This hairy mass often extends more than half the length of the arms. These two characteristics make *E. mirabilis* easily distinguishable from other *Espostoas* in the region. (Fig. 2.198, 2.199, and 2.200.)

E. calva is a yellow-spined columnar that grows on steep hillsides above the Río Marañon.[128] Despite my two visits to the region, I have not managed to photograph it.

E. laticornua is a handsome, very large, candelabra-shaped cactus common in the vicinity of Bagua and near Jaen and along the Río Chamayo in Peru. It may be a synonym of *E. lanata.* (Fig. 2.201.)

E. utcubambensis appears on the scene most strikingly between 1000 m and 2200 m elevation along the upper Río Utcubamba, a wild, crashing tributary of the Río Marañon that runs parallel to it for more than 100 km. The canyon of the Utcubamba is narrow, supporting villages at only a few places where side canyons create deltas and buildable

Figure 2.198.
Espostoa
blossfeldiorum,
Río Marañon,
Peru. Note brushy
skirt at the base.
Also in the photo
are Armatocereus
rauhii *and*
Browningia
pilleifera.

Figure 2.199. Espostoa mirabilis, *near Balsas, Río*
Marañon, Peru.

Figure 2.200. Cephalia of Espostoa mirabilis.

Figure 2.201. Espostoa *cf.* laticornua *near Jaen, Peru.*

Figure 2.202. Espostoa utcubambensis *near Tingo, Río Utcubamba, Peru.*

terraces along with them. These large cacti—thousands of them, locally called *tununas*—grow for many hundreds of feet above the river, towering over the vegetation that remains on the steep hillsides (clearing for pasture and crops has decimated most of the original forest, even on very steep slopes), but toward the bottom the great plants compete with tropical trees for sunlight and thus grow quite tall. Many of these very numerous cacti exceed 10 m in height, often sporting orchids and bromeliads on their branches and trunks. Several people with whom I spoke remarked on the usefulness of the woolly cephalium, but no one knew of anyone who still stuffs pillows with it. (Fig. 2.202.)

Northwest across the river from Bagua Chica, a sultry city on the Río Marañon, one begins to see very large, spreading *Espostoa.* As one approaches the agricultural city of Jaen, *E. superba* plants become more and more numerous, sharing the spotlight with fine examples of *E. laticornua,* of *Armatocereus rauhii* that grow in single stalks like totem poles, and of the spreading, tall *Browningia altissima.* The giant *superbas* sport brilliant white lateral cephalia. Seen as single, spreading trees, they are handsome indeed. I wish I could have one in my yard, but they are distinctly tropical plants, and the very thought of frost, which we usually get in Tucson ten to twenty times each winter, would kill the poor plants. Charles suggests that *E. superba* is probably a synonym of *E. lanata.*[129] It is found around and north of the city of Jaen.

I supposed that these marvelous cacti would surely be of ethnobotanical interest to someone, but found no takers. The region in which they are found is cocoa- and pineapple-raising country, which also sports vast acreages of rice and many mango and papaya trees. Who would need cactus fruits in such a cornucopia of large, juicy mangos, papayas, and pineapples?

The Tall Ones of Peru and Northern Chile: *Neoraimondia, Weberbauerocereus, Corryocactus,* and *Eulychnia*

I consider these genera together for no other reason than that they are hardcore dwellers of the desert and seem to have evolved under conditions drier than those of any columnar cactus of North America, with the possible exception of parts of Baja California. For the most part, they populate western Peru and northern Chile.

Neoraimondia

These strange, large cacti seem anatomically distinct from all others. Although taxonomic controversy surrounded the genus *Neoraimondia* until recently (as many as five species have been claimed), most authorities now agree that there are but two species, both massive columnars, one from Peru, the other from Bolivia.

Neoraimondia herzogiana (Backeberg) Buxbaum
caripari

If one cactus is notable by the very slight attention it receives in this book, it is this giant, the tallest and perhaps the most massive of all South American columnars. Commonly called *caripari,* it grows in the lower central valleys of Bolivia, where it reportedly reaches nearly 15 m in height—almost 50 ft. The fruits are renowned for their sweetness, and the plants produce bountiful harvests. (See fig. 2.1.)

Neoraimondia arequipensis (Meyen) Backeberg
sapang haurni (Quechua)

N. arequipensis, without a central trunk but with branches that may reach 30 cm in diameter, grows in the very dry desert of western Peru, where widely scattered individuals may be the only visible plants growing on the starkly barren hillsides. It grows as low as sea level, lower than the *Browningia candelaris,* with which it competes for the prize of being able to survive with the least rainfall. So distinct is its appearance that confusing it with any other cactus is nigh unto impossible. (Fig. 2.203.)

The thick, succulent stems of *N. arequipensis* seem oddly out of place in such an environment of nearly total aridity. The plant has four to eight ribs, which may be raised several centimeters from the central axis, so that in cross section the plant appears to be mostly rib. This succulence renders the plant capable of absorbing huge amounts of water on those unusual occasions when it rains, and it becomes more round than ray shaped in cross section. Where rain, though sparse, is reliable, the cactus forms groves (for example, near

Figure 2.203. Neoraimondia arequipensis, *Cañon Tinajas near Lima in central Peru. These large plants seem capable of surviving with no rainfall. El Niño rains turned this landscape green.*

Villenga in Cañon Cotahuasi) with numerous individuals exceeding 7 m in height. Some 20 km north of Aplao (in the western drainage of the Río Majes canyon) at about 2000 m elevation, the bajadas of the steep eastern slopes of the valley support a fine population of *sapang haurnis* (the Quechua name, meaning "lonely woman") that reach a height of more than 7 m. (Fig. 2.204.)

Neoraimondias vary greatly in habit, height, arm diameter, and even color. In many (but not all) plants, the stout, strong, potent spines may reach more than 20 cm in length. Though the spine numbers per plant may be limited, the sheer length and strength of the spines can be intimidating, indeed. Pre-Columbian peoples formerly used (and their ancestors to some extent still use) them as a tool for sewing and prying. They are called *quiscas* in Quechua, and archaeologists have discovered them in ancient sites near Nazca, Peru.[130]

Wherever the fruits grow, they are eaten by natives and by birds, whoever gets to them first. In villages of the middle Río Majes, the fruits may be added to the fruity concoction fermented into *chicha,* the national native brew of Peru. In much of *Neoraimondia* habitat, human beings are distant, for freshwater is utterly lacking, so animals are the beneficiaries of fruit production. Ostolaza, Mitich, and

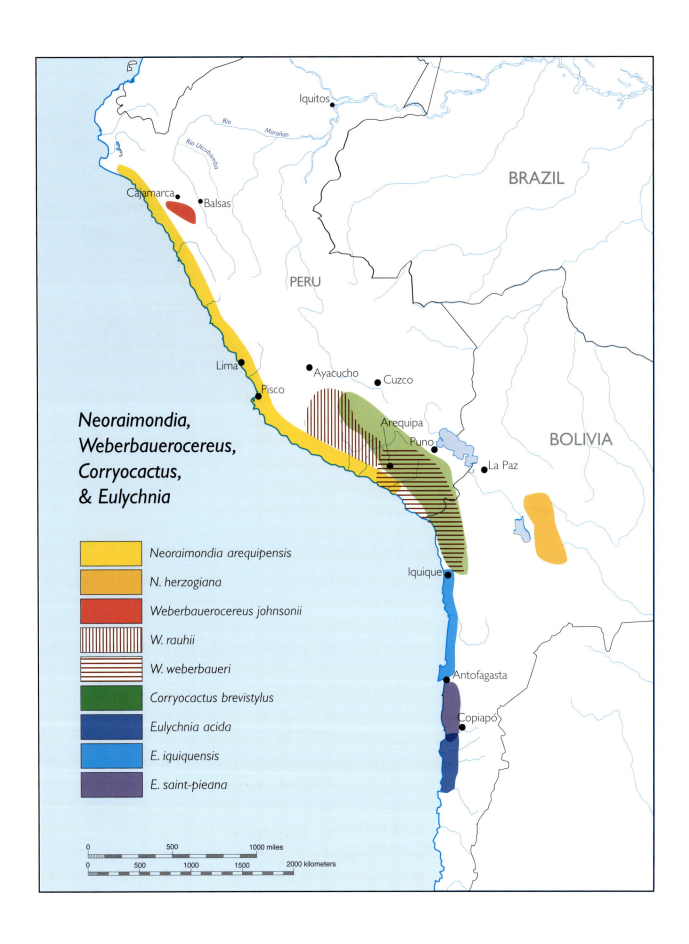

Neoraimondia,
Weberbauerocereus,
Corryocactus,
& Eulychnia

Neoraimondia arequipensis

N. herzogiana

Weberbauerocereus johnsonii

W. rauhii

W. weberbaueri

Corryocactus brevistylus

Eulychnia acida

E. iquiquensis

E. saint-pieana

Figure 2.204. Neoraimondia arequipensis, *Río Jequetepeque valley, northern Peru.*

Figure 2.205. Neoraimondia arequipensis, *Tiabaya, Arequipa, southern Peru.*

King note (without citation) that ethnobotanist Richard Schultes found Peruvians who stewed pieces of the stems to produce a liquid called *cimora,* which is commonly believed to have psychoactive properties.[131]

The wood is of variable texture. In some plants, it is of marginally useful durability; in others it is too soft to be of any use except for livestock fodder. It is a wonder that such weak wood can support these massive columns of water.

The most notable anatomical feature of this notable genus is its areoles. They continue to lengthen and broaden, and they flower repeatedly throughout the life of the plant, which, according to natives, may be more than one hundred years. As a result, the areoles grow grotesquely large, all out of proportion to the areoles on other columnars, at times resembling enormous nipples up to 5 cm in diameter and 8 cm in length. Another odd characteristic is the asymmetrical growth pattern whereby plants in very dry parts of the genus's range may devote all their energy into the growth of one stem or branch, leaving the others dormant. Thus, it is not unusual to find widely varying conditions among the branches, some appearing green and shiny new, others appearing old and decrepit. (Figs. 2.205, 2.206, and 2.207).

N. arequipensis plants do not grow into candelabra form, for they appear to lack a detectable central trunk. Instead, the stalks (or multiple trunks) may be one or many, emerging from the ground but branching only infrequently. *N. herzogiana,* like *Browningias,* develops a strong trunk and assumes an unambiguous candelabra form.

Near Tiabaya, Arequipa, a fine assemblage of cacti grows on the arid (less than 100 mm of rainfall per year) granitic hillsides. Here *Neoraimondias* intermingle with *Browningias* and *Weberbauerocereus weberbaueri* in a rich association of cacti with little competition from other plants. Indeed, the flora of the Department of Arequipa hosts more species of cacti than any other plant family. Among all these cacti, the *Neoraimondias* stand out for their almost gross succulence.

Weberbauerocereus

***Weberbauerocereus weberbaueri* (Vaup) Backeberg**

***Weberbauerocereus* sp.**

***W. rauhii* Backeberg**

***W. johnsonii* Ritt.**

Among the many columnar cacti of Peru is this somewhat obscure, very spiny, and rather nondescript genus. Its general aspect and flowering patterns suggest that lump-

Figure 2.206. Neoraimondia arequipensis *near Aplao, Río Majes, Peru. Local residents add the fruits to chicha, a mildly alcoholic drink made from corn.*

Figure 2.207. Areoles of Neoraimondia arequipensis. *Unlike other cacti, these areoles flower repeatedly and continue to grow in length.*

ers will sooner or later assume the genus into the massive group now labeled *Echinopsis*. Two of the species described here, *Weberbauerocereus weberbaueri* and an apparently undescribed species from the lower canyons (below 2100 m elevation) of southwestern Peru, are abundant in Arequipa, flourishing in the often startling aridity of the steep, barren slopes where annual rainfall is less than 100 mm. The flowers of *W. weberbaueri* are unusual in that their color changes from a bright green-yellow upon opening to a delicate purple-pink on maturing. Thus, it is not unusual to see flowers of different colors on the same plant, raising the possibility that the plants can be pollinated by both bats and hummingbirds.[132] This species can be shrubby or columnar, depending on the substrate and, perhaps, the rainfall. Near the village of Tiabaya, the plants are shrubby, but reach 5 m in height in association with *Neoraimondia arequipensis* and *Browningia candelaris*. The fruits are slightly sweet and edible, though somewhat mealy. A local woman reported that they are eaten when sour (still green) during the "cambio de vida" (change of life, menopause) to help calm the consumer and alleviate her discomfort. The local name for the plants is *huarango*, a generic term in the region for "large shrubby plant." They are found in large numbers on very dry slopes between 2200 and 2800 m. On the hill called Yarabamba near Arequipa, treelike specimens with dozens of branches and fourteen to sixteen ribs seem to cover the hillsides. Annual

rainfall of less than 100 mm does not seem to deter them. (Figs. 2.208, 2.209, 2.210, and 2.211.)

Recently germinated plants are easy prey for goats, so juvenile plants are uncommon in the vicinity of human habitation. The dry wood can be used for kindling, but is rather soft and lacking in substance.

An apparent second species, perhaps undescribed, is more strongly columnar, reaching greater than 7 m in height, and is more compact and vertical than *W. weberbaueri*. It grows in the Department of Arequipa on lower slopes, below 2100 m elevation and down as low as 1600 m on very dry hillsides. Its habit and spinal coloration suggest that it is a different species from *W. weberbaueri*, but its description does not correspond with anything I have found in the literature. It has sixteen to twenty ribs, compared with fourteen to sixteen for the plants of Tiabaya and environs. Near Judiopampa, below the cataract of Sipia on the Río Cotahuasi, plants as tall as 9 m grow in dense groves in association with *Armatocereus riomajensis*. Residents of the region report that the wood from dead plants is resistant and durable, strong enough for braces and cross pieces in looms. The fruits, however, are of indifferent flavor and thus ignored. The cactus is called *huasa sanquey* in the region. (Fig. 2.212.)

A third species, apparently *W. rauhii*, frequents the steep, granite slopes of the canyon of the Río Pisco above 1800

Figure 2.208. Flower of Weberbauerocereus weberbaueri *near Arequipa, Peru. Plants may support both yellow and pink flowers.*

———

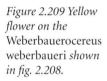

Figure 2.209 Yellow flower on the Weberbauerocereus weberbaueri *shown in fig. 2.208.*

Figure 2.210. Large Weberbauerocereus weberbaueri *near Yarabamba, Arequipa, Peru.*

———————————

Figure 2.211. Large Weberbauerocereus weberbaueri *on hillsides near Yara-bamba, Arequipa, Peru.*

Figure 2.212. Weberbauerocereus *cf.* rauhii, *Cañon Cotahuasi, Peru.*

Figure 2.213. Weberbauerocereus *cf.* rauhii *near Huanterá, Huancavelica, Peru.*

m elevation, where it accompanies *Armatocereus rioma-jensis* and *Neoraimondia arequipensis.* Natives refer to this many-branched tree with more than twenty ribs on the branches as *lanzón,* attributing liver-benefiting properties to the pulp. (Fig. 2.213.)

Also tall and columnar is *W. johnsonii* of northern Peru. Never especially abundant, this worthy plant reaches nearly 7 m in height. It grows in rather moist mountainsides above the valley of the Río Jequetepeque, but its habitat has been severely compromised by clearing for agriculture, and large plants are now quite rare.[133] I have not seen it, and its days may be numbered.

Corryocactus

Corryocactus brevistylus (K. Schumann) Britton & Rose
sancayo

Corryocactus sp.

Sancayos produce the largest fruits of all wild columnar cacti, perhaps of any wild cacti. The only competitor is

Stenocereus gummosus. A large fruit weighs in excess of 250 g (half a pound) and is as large as a softball. The fruits are gathered in hill country to the south of Arequipa and hawked in the city's open-air markets in March through May each year. When free of spines (the rather weak spines slough off easily), the fruits resemble large, ripe, round avocados. The oddest feature of the fruits is that although one might expect a toothsome gob of sweet pulp, the attractive greenish pulp is sour, even acid to the taste. I am accustomed to eating the sweet, juicy pitayas of Mexico, and although I was warned by a vendor that the fruits were not sweet, I was surprised and, I confess, not a little taken aback at my first taste, for it was tart indeed. Several onlookers found my enthusiastic initial consumption quite amusing. (Figs. 2.214 and 2.215.)

It comes as no surprise, then, that the most prominent virtue according to those who use the fruits is not their properties as food, but their medicinal qualities. When I asked people what the fruits are good for, they invariably answered, "They are good for the liver." They would often add, "They are good for the kidneys, too." When I persisted, they would explain that eating the fruit on an

Figure 2.214. Sancayera (Corryocactus brevistylus *fruit seller*) *in market, Arequipa, Peru.*

Figure 2.215. Sancayo (Corryocactus brevistylus) *fruit.*

empty stomach would "cleanse" the liver, removing impurities and strengthening its function. The same would go for the kidneys. So the fruits are believed to be a remedy for dirty kidneys and livers. Belief in the salubrious effects of columnar cacti fruit seems to pervade Peru and Bolivia.

The fruits are eaten for their own sake, however, which is a good thing, for in their range the plants become common indeed, and each may produce several dozen fruits.[134] Arequipans add sections of the pulp to salads, cutting the powerful acidity with limejuice. In the vicinity of the Quechua town of Chivay in Cañon Colca region, natives prepare a mixed alcoholic drink called the Colca Sour, blending sancayo fruit, lime juice, sugar, and rum.

Several older women also mentioned that the sap from the trunks makes a high-quality shampoo. Few people use it nowadays, they all lament, because it is easier to purchase shampoo than to make one's own from the sancayo. In Chiguata, an area known for an abundance of sancayos, residents mentioned that the dried wood makes for decent cooking wood (no house in the entire region has gas heating, and few have gas for cooking).

C. brevistylus is found from the Arequipa region southwards into Chile. I have seen what I believe to be sancayos on the slopes of the Cañon Pisco at an elevation of 2500 m as well. The flowers are bright yellow, blooming in April, perhaps in other months as well. It flourishes at this elevation and grows well above 3000 m. It seems to require more rainfall than many other Peruvian columnars, perhaps flourishing where rainfall exceeds 200 mm. It appears to be especially common along ancient rock terraces, which are extraordinarily abundant reminders of human achievements of old throughout much of montane southern Peru, following hillside contours over thousands of square kilometers. The terraces concentrate warmth and rainfall, perhaps enhancing the sancayos' natural habitat and assisting in the selection of varieties capable of better production of both fruit and firewood. (Fuel woods are scarce in the dry and high-elevation western slopes of the Andes.) Plants in the Arequipa region reach little more than 3 m tall and are armed with wiry, straight spines up to 10 cm long.

In Cañon Colca, north of the city of Arequipa, a race (or, more probably, a separate species) of sancayos is to be found that rivals several species of *Stenocereus* in size and habit. It is uniformly upright and treelike, in contrast to the race growing south of Arequipa, which tends to be shrubby at times. The plants grow in dense aggregation along the steep canyon slopes and reach elevations of nearly 4000 m. Some plants exceed 8 m in height or more and are less viciously armed than those in the Arequipa region. They

Figure 2.216. Sancayos growing on ancient terraces, Cañon Colca, Peru.

are especially common on hillsides too steep for agriculture or terracing, but where humans have been active for thousands of years, suggesting that a result benefiting both species has occurred. They form forests at certain locations in the canyon. They often are host to numerous ballmosses and other bromeliads, which lend them a ratty appearance. (Fig. 2.216.)

A similar, but more treelike form or separate species appears on steep hillsides in Cañon Pisco. An apparently distinct species, noncolumnar or only semicolumnar, grows on cliff faces and very steep hillsides of Cañon Colca, spreading out recumbently, branches occasionally dangling hooklike from roots attached to sheer rock faces. (Figs. 2.217 and 2.218.)

Corryocactus is evolutionarily obscure, derived from different pathways than those of the Browningieae and Trichocereeae, perhaps more primitive. Wallace, using cladistic studies, suggests that it is a separate, more primi-

tive clade, closer to the proto-cactus from which the other species are derived.[135] In appearance, though, especially the plants in Cañon Colca, it closely resembles other candelabra-shaped columnar cacti.

Eulychnias—*Children of the Fog*

Eulychnia acida Philippi 1864

E. iquiquensis (K. Schumann) Britton & Rose

E. saint-pieana F. Ritter

Eulychnias are generally classified as belonging to Notocacteae, a group for the most part not columnar in habit. These remarkable plants have somehow evolved into legitimate columnarhood, in spite of their growing in areas where it seldom or never rains. They thrive or at least endure in habitats where heavy fog, sufficient to produce condensation, is reliable for the better part of the year.

Figure 2.217. Corryocactus *cf.* brevistylus, *Cañon Pisco, Peru.*

Figure 2.218.
Corryocactus *sp., Cañon Colca, Peru. Note the dangling branches.*

That any terrestrial plant can survive without rainfall seems miraculous, but that a cactus can grow to be several meters in height in the complete absence of rainfall seems downright astonishing.

Only one species, *E. ritteri,* is found outside Chile. It grows in extreme southern Peru and may not make it into Chile. The others are Chilean endemics, confined to that part of the coastal and near-coastal Atacama Desert that is regularly visited by fog.

Eulychnias (most have no widely used common name) generally prefer rocky slopes where fog and mist condense on the plant branches and nearby rock and trickle down to the roots. At times, some plants can be found on gentler slopes and even flats, but usually only where there is some rainfall. Even in these sites, survival is remarkable, for in the *Eulychnias'* range almost no location receives more than roughly 50 mm of rainfall per year. The most dramatic plants are those growing on granite slopes near

Figure 2.219. Color photo of Eulychnia iquiquensis *on granite slope near Antofagasta, Chile. It never rains in this location, so plants survive on moisture derived from dew. Although they appear dead and ugly, they survive rather well.*

Figure 2.220. Eulychnia acida *near Copiapó, Chile. Average rainfall in this location is less than 50 mm (2 in.).*

Figure 2.221. Eulychnia saint-pieana, *Llanos de Challe National Park, Chile.*

Antofagasta, where rain has not fallen since time immemorial. (Fig. 2.219.)

For the most part, *Eulychnias* can hardly be called handsome plants, and those near Antofagasta are certainly no exception. Their growth is irregular, and their constant struggle for moisture must take its toll on any fancies the plants might have toward symmetrical or regular development, so many of them are raggedy and sometimes downright ugly. The lack of rainfall results in a pallid gray color, sometimes downright ugly. Most of the plants become covered with grime and dust that is never removed by cleansing rain, and they resemble drab fossils more than living plants. Parts of the plants are usually dead or dying, and the living parts may be dormant, so that it is difficult to ascertain which is which. Some examples of *Eulychnia acida* farther south, however, appear to be robust and growing, perhaps because they are able to supplement their regime of fog moisture with some rainfall. (Fig. 2.220.)

Of the four *Eulychnia* species (perhaps only three will withstand taxonomic scrutiny), only one has a widely

Figure 2.222. Eulychnia saint-pieana, *Llanos de Challe National Park, Chile.*

accepted common name: *Eulychnia acida* is widely known as *copao.* If the others do have common names, I was unable to ascertain them, even from naturalists at Pan de Azúcar and Llanos de Challe National Parks. Some people refer to them as *quisco,* but this term may merely be local. (Figs. 2.221 and 2.222; see also fig. 1.23.)

The San Pedro Cactus

Echinopsis pachanoi (Britton & Rose) Friederich & G. D. Rowley
ahuacullo, achuma, San Pedro

The *ahuacullo,* or San Pedro, is easily the best-known and most widespread columnar cactus of Ecuador and Peru, perhaps of all the Andes. (It is called *achuma* or *huachuma* farther south in the Andes.) It is found in its native state in southern Ecuador between 1500 and 3000 m elevation,

where it is locally common on steep banks, hillsides, and cliffs above watercourses. It ranges south through Peru and into Bolivia. It is also widely cultivated, gracing gardens, yards, fencerows, and parks throughout Ecuador and the outer world as well.[136] I have a fine specimen (that does not yet flower) more than 2 m tall in my yard in Tucson, Arizona. Information about every possible aspect of the plant is available from an abundance of Internet sites. (Fig. 2.223.)

The ahuacullo is a stately and handsome cactus, with robust plants reaching 8 m in height, although such tall plants are uncommon. Many horticultural plants lack spines and are thus common in pots, near doorways, and in gardens. The plants grow rapidly from cuttings, perhaps as much as one meter or more per year. The seven-ribbed columns are erect and tall, and the cactus gives a rather smooth appearance, with a blue-green hint of color. The flowers are white and quite large, up to 20 cm wide. They

Figure 2.223. Cultivated San Pedro cactus (Echinopsis pachanoi), *Vilcabamba, Ecuador. The plants are popular for landscaping, though consuming them is illegal in Ecuador.*

Figure 2.224. San Pedro cactus flower, Paracas, Peru.

Figure 2.225. San Pedro cactus fruit. Like several other South American columnar cacti, it bears scales instead of spines.

open shortly after sundown and are apparently pollinated by insects, although bats are quite probably involved as well.[137] The fruits are not juicy and apparently are not widely consumed, and I have been unable to document anyone who can testify to their flavor or lack thereof, or whether they concentrate any weird chemicals. (Figs. 2.224 and 2.225.)

Echinopsis pachanoi is the only columnar cactus widely known to be cultivated specifically for its hallucinogenic properties.[138] Several other species are known to contain mescaline, and *Trichocereus cuzcoensis* has a whispered reputation as a source of mind-altering chemicals, but to my knowledge no others are specifically cultivated for the high they are believed to deliver. All parts of the San Pedro appear to concentrate mescaline, the alkaloid that is significantly concentrated in peyote *(Lophophora williamsii)* as well, a property discovered untold millennia ago by early Peruvians and passed on to the Incas. Many people attest that when they ingest the mescaline, they experience visions of unmatched grandeur. The *Encyclopedia of Psychoactive Substances* reports the following:

> Like many other of the entheogenic substances used in the aboriginal religions of the Americas, the use of the hallucinogenic San Pedro cactus is ancient and its use has been a continuous tradition in Peru for over 3,000 years. The earliest depiction of the cactus is a carving, which shows a mythological being holding the San Pedro. It belongs to the Chavín culture (c. 1400–400 BC) and was found in an old temple at Chavín de Huántar in the northern highlands of Peru, and dates about 1300 BC. A particularly surpris-

ing discovery was made by a Peruvian archaeologist named Rosa Fung in a pile of ancient refuse at the Chavín site of Las Aldas near Casma; namely what seem to be remnants of cigars made from the cactus. Artistic renderings of it also appear on later Chavín artifacts such as textiles and pottery (ranging from about 700–500 BC). The San Pedro is also a decorative motif of later Peruvian ceramic traditions, such as the Salinar style (c. 400–200 BC), and the Nazca urns (c. 100 BC–AD 700). It has also been proposed that a recurrent snail motif in Moche art represents a mescaline-soaked snail, which has partaken of the San Pedro. If this is the case then the snail may be added to the list of animals having psychoactive properties.[139]

Legend has it that the Christian priests sent into the Andes by the conquering Spaniards condemned the ceremonial use of the cactus and forbade its cultivation. The possibility of uncontrolled and autistic visions does not mesh well with orthodox religion. (Wine was acceptable only for communion and for the clergy.)[140] The indigenous people responded by honoring the cactus with the name *San Pedro,* hoping that by invoking the name of the supposed founder of Catholicism, they might assuage the clerics' stern opposition to the ritual consumption of the plant. Another explanation for the name is that the cactus was like St. Peter, authorizing admission to heavenly visions.

Whether the priests were thus moved to soften their opposition to consumption of the cactus, its widespread use persisted and continues today, so much so that the plant can be viewed as semidomesticated, and apparently

Figure 2.226. Sections of San Pedro cactus intended for ceremonial use being sold in the municipal market, Chiclayo, northern Peru. In Peru, consumption of the cactus is legal and popular.

wild individuals are undoubtedly genetic mixtures of wild and cultivated parentage. Vilcabamba, in Ecuador's Loja province, is widely known as a center of San Pedro consumption, mostly by outsiders. I inquired there into the techniques for preparation, but received a chilly reception to my queries. Local peasants explained that consumption of the cactus is highly illegal in Ecuador, and those caught preparing the cactus have been imprisoned for up to six years. Peasants complain that the police in the Vilcabamba area are especially aggressive at weeding out imbibers. Nevertheless, others told me that many people, including natives, regularly consume the cactus and seem none the worse for having subjected themselves to the mind-altering experience. Firsthand testimony is hard to come by, however. Tyrannical governments disdain substances that produce altered states of mind, except for those favored by the governing few, and try rigorously to control them. Fortunately, the Internet is rife with personal accounts. On the skeptical side, noted Peruvian cactologist Dr. Carlos Ostolaza has repeatedly ingested the San Pedro decoction in ceremonies and reports that he has yet to experience a mescaline high.[141]

Peasants near Vilcabamba also reported that *San Pedro* is actually the name of the hallucinogenic potion that is consumed to induce visions, whereas the cactus itself is called *ahuacullo*. Although I respect the honesty of their report, the fact is that outside of the region where it grows wild, the cactus itself is nearly universally referred to as the San Pedro, so I perpetuate that name in this book. The same peasants noted that cows eat the cactus from time to time

and become disoriented and goofy as a result. The fellows found this effect to be most amusing, but I suspected it was not *their* cows that had eaten the cactus. Still, the San Pedro's and other hallucinogenic plants' effect on herbivores is probably precisely what the plants had in mind. If an herbivore is foolish enough to eat them, its behavior will be altered, probably to its detriment.

If Ecuador has repressive laws governing the San Pedro cactus, Peru is remarkably open. Not only is consumption perfectly legal, but practitioners abound who supervise the preparation and administration of the San Pedro potion, especially in Peru's northwest. A saleswoman in the Chiclayo market described for me its preparation. An indeterminate chunk of the flesh is boiled with water to dissolve the chemical contents of the plant. The resulting liquid is skimmed to remove impurities and then simmered until it becomes somewhat concentrated. It should then be mixed with some cane liquor *(aguardiente)* and drunk in small doses. Taken thus, it will not produce nausea, she assured me. In the absence of the aguardiente, the liquid can be drunk straight, but others have mentioned that it tastes simply awful. Statuary and ceramics from Chavín de Huántar, Peru, portray San Pedro users with streams of mucus running from their noses—a good sign, from their perspective.[142] Most users consider the trip produced by the psychoactive components well worth the bad taste and the frequent nausea produced by the ingestion. They describe visions of swirling color and clear perceptions of previously hazy realities. Some day I may try it. (Figs. 2.226 and 2.227.)

Historically, however, the brew was prepared by *curanderos* (healers) or shamans and administered under more or less carefully controlled conditions.[143] In Peru, it is administered (or requested) as an alternative treatment for illnesses that do not respond to conventional remedies, to improve the imbiber's fortune, and to maintain family harmony. Some curanderos report that the potion is a purgative and purifier.[144] The onset of hallucinations is not an essential part of the curing ceremony, but appears to be a welcome benefit when it occurs. In the city of Chiclayo, situated in a zone rich with remnants of pre-Incan urban and religious sites, the market hosts a section devoted to *brujería* (witchcraft). In several stalls, cuttings of the San Pedro cactus are stacked, awaiting purchasers who will transform them into a hallucinogenic potion. For those wishing to eschew the elaborate preparation through boiling, a powdered form is also available.

The San Pedro is more willing to sprout from cuttings than almost all other columnar cacti, except for such eager sprouters as *Cereus hildmannianus* and *Pachycereus [Lophocereus] schottii*. Horticulturalists widely report that ahuacullo readily accepts grafts of other cactus species. One wag reported attempting to graft peyote buds onto stems of the ahuacullo, thus producing a supermescal concentrator. No results have been publicized so far.

San Pedro joins a distinguished list of hallucinogenic plants from the Andes. In addition to the celebrated *ayahuasca* and coca leaf, the powder of the ground dried seeds of *cebil* (the leguminous tree *Anadenanthera colubrina* of northwest Argentina, Paraguay, and Bolivia) is now known to be capable of inducing powerfully altered states of consciousness. It was widely popular among pre-Hispanic Andean cultures. Argentine authorities might look askance at the plant, but it is a very handsome and common large tree, flourishing in a variety of moist to semiarid habitats. It is also a pioneer plant in disturbed soils, sprouting in large numbers on cutover hillsides. Eradication would be expensive and futile, as it would be for the ahuacullo.

Ecuador is home to at least seven additional columnar cacti, but none of them figures prominently in local uses. This seems a shame until one realizes that Ecuador is also home to an astonishing variety of wild plants and amenable to raising a huge variety of orchard plants as well. Nearly all of the country was once covered with forests of staggering biological diversity, including thornscrub and rich tropical deciduous forests on the coast and the lower slopes of the coastal hills; semideciduous tropical forests; dripping montane cloud forests at midelevations; complex forests that have at least three species of the coniferous *Podocarpus* at higher elevations in the Andes; and, crossing

Figure 2.227. San Pedro cactus in flower, Cuenca, Ecuador.

the Andes, the incomparable lowland rain forests of the Amazon. The variation in habitats is equally varied, often changing dramatically in only a few kilometers as cloud traps give way to rain shadows that merge into fog belts in unpredicted convolutions. Verdant valleys turn sere at nearly the same elevation, only to turn green again at a slightly lower elevation. Prediction of plant communities simply based on elevation or orientation is difficult. Most surprising to the cactologist, however, is the appearance of arid bands with desert vegetation in several southwestern valleys. Even humid Ecuador has its deserts.

The Sausage Cacti: *Armatocereus* and *Jasminocereus*

Armatocereus

Armatocereus, the most geographically widespread genus of Ecuador and Peru, with one species appearing in Colombia, appears to be primitive or, perhaps more accurately, less derived from ancestral cacti than are most other columnar

Figure 2.228. Suja (Armatocereus riomajensis), *Mollebaya, Arequipa, Peru.*

cacti. It consists of thirteen or so species, but some of the species have been insufficiently studied by taxonomists to fix the number with precision. Some researchers now group *Armatocereus* with *Leptocereus* as more or less direct descendents of primordial cacti.[145] Comparing plants in this genus with the primitive *Calymmanthium* of the Río Marañon region in Peru, one can almost sense that the latter tall, meandering, branchy cactus struggling for columnar-hood is a progenitor of the *Armatocereus*. Furthermore, within the *Armatocerei,* the species range from almost bushy to spreading to elegantly upright. Members of the genus possess an easily identified habit, perhaps more so than any other genus: they resemble strings of sausages. (Fig. 2.228.)

Most if not all Armatos have white flowers that open at night. Fruits of most species have white pulp and black seeds. Some of the species (especially *A. rauhii*) are very attractive and should make handsome additions to cactus

Figure 2.229. Edible suja fruit.

gardens. I suspect that due to the equatorial evolution of the genus, most species will be very cold sensitive. (Fig. 2.229.)

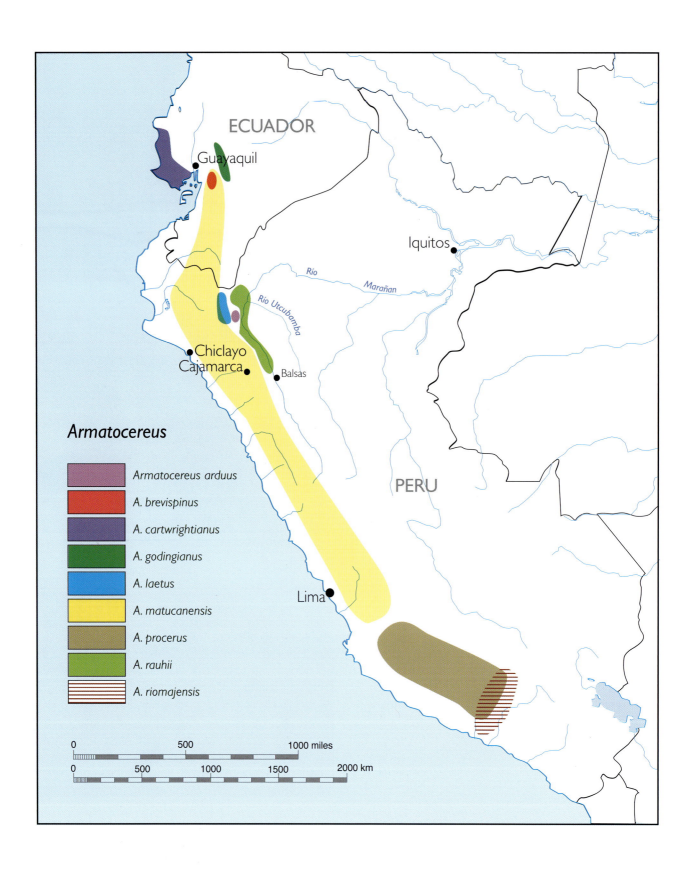

ECUADOR

Guayaquil

Iquitos

Rio

Marañan

Rio Utcubamba

Chiclayo
Cajamarca

Balsas

PERU

Lima

Armatocereus

	Armatocereus arduus
	A. brevispinus
	A. cartwrightianus
	A. godingianus
	A. laetus
	A. matucanensis
	A. procerus
	A. rauhii
	A. riomajensis

0		500		1000 miles

0	500	1000	1500	2000 km

It is odd that South American common names for the great columnar cacti are singularly drab. For example, on the coast of Ecuador, where *Pilosocereus tweedyanus* and *Armatocereus cartwrightianus* are common and evident, both are referred to simply as *cactus.* The species *A. cartwrightianus* is also called *cardón,* as might be predicted.[146] In Peru, the sausage cacti are commonly referred to as *gigantes,* or giants. I tried in vain to locate residents with knowledge of pre-Hispanic languages who might recall indigenous names and could find only the generic Quechua term for columnar cacti, *sanquey.* Apparently, when native cultures were exterminated, local plant names were lost along with them.

This lack of interesting common local names is a shame because the cacti in this area are worth noting, especially the attractive, varied, and sometimes handsome members of the genus *Armatocereus.* Four species of this genus can be found in a relatively small area in Ecuador, so that all can be seen in one very long day by motorcar. Nine or so are to be found in Peru, but are rather widely spread over the length of the country. All members of the genus feature segmented branches that lend the plant a characteristic appearance and lead me to call them *sausage cacti.*

First, the Ecuadorans.

Armatocereus brevispinus J. E. Madsen

This very rare plant is a large (up to 10 m tall), many-armed columnar cactus. It is apparently endemic to Ecuador[147] and confined there to steep hillsides at elevations 1700–2000 m around the semiarid Catamayo Valley in what was probably tropical deciduous forest before being pillaged and cleared. The individuals that remain flourish, but only in areas inaccessible to agriculture and livestock and not much more easily accessible to biologists: they are invariably embedded within other dense thickets of shrubs and small trees. *A. brevispinus* is a handsome plant, unlike the often scrofulous-looking *A. cartwrightianus* of the lowlands. Its isolation and tendency to grow amid vigorously flourishing low evergreen forests makes it difficult both to photograph and to imagine how a plant would appear standing alone. Although this plant must be considered extremely rare, its preference for steep slopes among thorny vegetation bodes well for recruitment of seedlings and perpetuation of the species.

Armatocereus cartwrightianus (Britton & Rose) Backeberg ex A. W. Hill
cactus

A. cartwrightianus is the most common member of the genus in Ecuador. It grows abundantly in the arid to semi-

Figure 2.230. Armatocereus cartwrightianus, *Machalilla National Park, Ecuador.*

arid lowlands over most of southwestern Ecuador. It can hardly be called a handsome plant, for although the upper branches are green, the lower arms and branches usually turn gray, and the old bark tends to slough off, giving the plant a sickly or moribund appearance. The arms grow erratically in sausagelike segments, each sausage representing one year's growth.[148] An arm does not necessarily add a new segment each year, however, so counting the continuous sausages in a string gives no hint of the plant's age. Cuttings or fallen arms tend to sprout new growth and may grow rapidly, but the growth is not uniform. The plants do endure for many years: residents of Machalilla National Park maintain that a plant more than 11 m high growing near Agua Blanca, an indigenous community with pre-Columbian roots, is more than eighty years old. (Fig. 2.230.)

At Agua Blanca, residents refer to *A. cartwrightianus* merely as *cactus* in spite of their indigenous heritage. Still, they collect the fruits, pronouncing them to have excellent flavor. They also mix the fruits (well beaten) with water and drink the juice, pronouncing it to be tasty and satisfying. In the dead cactus, the very hard skeleton is surrounded by spongy storage tissue that is soft and useful for a packing material in packsaddles. The wood itself is very hard, so much so that in Peru the plant's common name is *cardón madero,* or "the timber cactus."

Armatocereus godingianus (Britton & Rose) Backeberg ex E. J. Salisbury

Another endemic Ecuadorian member of the genus, this highland cactus is common in lofty valleys of south-central Ecuador, where it grows to nearly 10 m in height. It is noticeable along the precipice-like highway between Aluasí and the Río Chanchan in Chimborazo province. Whether the persistent snail-shaped fruits are eaten I was unable to determine. This species also is known locally only as *cactus.* I have no photographs because during my time in the region a vast, thick fog covered everything from the bottom of profound canyons to the tops of Andean peaks.

Armatocereus matucanensis Backeberg ex A. W. Hill

Uncommon, but more far ranging and much more accessible to inspection than *A. brevispinus* and more cosmopolitan than *A. godingianus,* is *A. matucanensis,* a rather shorter plant (a maximum 8 m tall) with many branches. It grows abundantly in the Catamayo Valley and on into southern Peru, colonizing low hills as well as the gradual and steeper slopes covered with short, thorny shrubs. Fine, spreading individuals prosper at 1600 m elevation in the mists of upper Cañon Tinajas near Lima and above 2400 m near Arequipa. Its more relaxed ecological requirements mean that it is better known to residents of its habitat, and its fruits are locally eaten. One might be discouraged from attempting to harvest the fruits, for they are covered with vicious-looking spines. When the fruit ripens, however, the spines slide off in an ersatz sleeve or jacket, leaving the ripe fruits unarmed and ready to eat, while on the ground below lies an artistically arranged airy spheroid of cactus spines, a phenomenon I have seen elsewhere only in the case of *Stenocereus chacalapensis* on the coast of Oaxaca, Mexico. Once again I was able to gather only the local name *cactus* for this admirable plant. Residents of the Catamayo Valley eat the fruits. In Peru, its fruits are widely regarded as delectable, but generally the plant grows in habitats far from historic habitations. In Cañon Tinajas, it becomes a

handsome plant known locally by the name *pitaya* (!). It grows into a spreading treelike plant with a massive trunk and hundreds of joints, none of them more than about 30 cm in length, together resembling a huge concatenation of sausages. (Figs. 2.231 and 2.232.)

The remaining cacti in this genus are endemic to Peru.

Armatocereus cf. *procerus* Rauh & Backeberg

On very steep desiccated slopes of granite inland from the Peruvian city of Pisco, these sausage cacti appear at about 1000 m elevation, the first columnar cacti seen journeying inland from the coast. They are hardly handsome plants since it almost never rains where they grow, and they are never washed. Under their cloak of dust, they are yellowish green and rise to more than 7 m. (Fig. 2.233.)

Armatocereus riomajensis Rauh & Backeberg sanquey (Quechua)

It is in Peru that *Armatocerei* strut their stuff, growing in a variety of habitats nearly into Chile (and perhaps there as well). In southern Peru west of the Andes, *A. riomajensis* is widespread, but especially notable at 2300–2800 m elevation on the gentler slopes in the profound declivities such as Cañon Colca, Cañon Cotahuasi, and Cañon Pisco. I found it also on fog-frequented slopes near Arequipa. It is locally referred to as *ampaca* or *sanquey* (Quechua for "columnar cacti"), not to be confused with the the name *sancayo* for *Corryocactus brevistylus,* often found growing in the same habitat.

A. riomajensis is typically taller than *A. matucanensis,* reaching more than 7 m in height, but less inclined to spread out. It is common between 2200 and 2500 m in southern Peru, where it flourishes on hillsides. The segments or sausages of this handsome species are also longer, and the plant is more nearly upright. The fruits, called *sujas,* are reportedly mostly reddish, but occasionally white. Native peoples and livestock gather them. In Cañon Cotahuasi, sanquey can be found in great numbers on the less steep hillsides arising from the canyon bottom. (Fig. 2.234.)

Armatocereus rauhii Backeberg gigante

Armatocereus cf. *laetus* (Kunth) Backeberg ex A. W. Hill

In northern Peru, at least three species of Armatos, and probably more, can be found in the Río Marañon region, a part of the Amazon basin that oddly enough flows generally northward *west* of the eastern spur of the Peruvian Andes

Figure 2.231. Pitaya (Armatocereus matucanensis), Cañon Tinajas, Lima, Peru. The purplish fruits are quite edible.

until it breaks through the shoulder of the great range and plunges northeastward. The most notable of these plants is *A. rauhii,* which is a handsome, erect string of sausages reaching upwards of 10 m in height with a distinct trunk. It becomes a large cactus of a characteristic pale green color. The more robust plants are fair indeed, forming attractive assemblages with other columnars, notably *Browningias* and *Espostoas.* A wondrous use of the plant emerged near Balsas: several residents of the region informed me in all earnestness that if a cross section of the plant is dropped into a bucket of muddy water, after a few hours the water will be purified, and all the mud will accumulate at the bottom around the cactus section. I had no time to check this operation—resident observers assured me that it was necessary to wait a minimum of two hours for the miracle to take place—but if true, it is surely worth investigating. The fruits of this fine plant, which is locally called *gigante,* or giant, are often eaten, but the wood is deemed too insubstantial to be of any use. In addition to the abundance of very large plants near Balsas, gigantes thrive along the Río Marañon at the point where it begins to flow northeast. Here the plants tend to have fewer arms, many of them being single stalks of a most agreeable appearance. This region is usually regarded as the hottest part of Peru, so *A. rauhii* should grow well in the Sonoran Desert with supplemental water. (Fig. 2.235.)

Figure 2.232. Armatocereus matucanensis, *Catamayo Valley, Ecuador.*

Figure 2.233. Armatocereus *sp., possibly* procerus, *Cañon Pisco, Peru.*

Figure 2.234.
Armatocereus
riomajensis, *Cañon*
Cotahuasi, Peru.

Figure 2.235. Armatocereus rauhii, *near Balsas, Río Marañon, Peru.*

Figure 2.236. Armatocereus laetus *near Tambo, Río Chamayo, Peru.*

Figure 2.237. Armatocereus *sp., perhaps* arduus, *Río Chamayo, Peru.*

West of Bagua a somewhat different Armato, one with no trunk, perhaps *A. laetus,* grows in steep, dry slopes of the Río Chamayo, a tributary of the Río Marañon that flows from west to east. This plant is as rambling as *A. rauhii* is upright. It reaches 6 m in height, but little more. I was unable to obtain any ethnobotanical information about the plant, largely because the habitat is only sparsely populated with humans. It appears that where these plants grow, rainfall is low and perhaps erratic, and many of the individual plants have lost branches or exhibit some sign of water stress. Anderson suggests that the species may not be distinct from the more widespread *A. matucanensis.*[149] (Fig. 2.236.)

Armatocereus cf. *arduus* F. Ritter

On the banks of the Río Chamayo, I came across a different plant growing in considerable numbers. I tentatively classify it as *Armatocereus arduus.* It is clearly rangier and far spinier than the more common *A. rauhii,* with a different habit entirely. The individual shown in figure 2.237 was growing among dense greenery in rich tropical thornscrub. I could obtain no ethnobotanical information about it. However, my experience suggests that it is probably called *gigante* by residents and is not distinguished locally from *A. rauhii.* (Fig. 2.237.)

Jasminocereus

Jasminocereus thouarsii (F. A. C. Weber) Backeberg candelabra

Of this strange and exotic-looking Galápagos Islands cactus one thing is certain: it has no aboriginal name, at least among extant peoples. Its Spanish name is *candelabra,* which seems inappropriate given its normally narrow habit. It is endemic to several islands in the Galápagos Archipelago. It exceeds 7 m in height on Santa Cruz Island and is usually very spiny. The spines on the lower parts of the trunk are especially thick, probably protection against the herbivory of Galápagos tortoises, the only plant-eating creatures on the islands capable of inflicting damage. These

Figure 2.238. Jasminocereus thouarsii, *Santa Cruz, Galápagos Islands, Ecuador.*

cacti cannot be said to be especially attractive plants, for they often appear rangy and partially dead, and even when tall they show enough gray and white to give them a less than wholesome appearance. However, when they appear in assemblage with tree prickly pears, *Opuntia echios,* as can be seen on Santa Cruz Island, they are attractive—indeed, part of an otherworldly landscape. Their resemblance to other sausage cacti is notable. (Fig. 2.238.)

I spoke with residents of San Cristóbal and Santa Cruz islands (the two most populated) about the plants. None of them was aware of any human uses. The fruits are the size of small plums, reddish, and with few or no spines. However, their endocarp is harder than orange rind and becomes even harder with maturity, making the pulp difficult to get at. Woodpecker Finches and (perhaps) Cactus Finches tear through this leathery skin and to their considerable advantage feast on the white pulp, which, although somewhat tart, is edible by humans as well. One

could eat several or, under duress, many. The seeds are tiny, similar to those of *Armatocereus* species or even to those of *Stenocereus thurberi* of North America. The fruits appear to remain attached to the plants through maturity, meaning that dispersal must be through the finches that eat the pulp. Naturalists with whom I spoke have not seen tortoises eating any fallen fruits, perhaps more of a reflection of the near extermination of the great tortoises than a conclusive observation on the natural history of the cactus.

I have not seen the flowers up close. Plants of the Santa Cruz Island subspecies (there are three subspecies, according to Anderson) are tall, and the flowers most inaccessible. Galápagos National Park regulations prohibit the removal of any flowers, so without stepladders that will permit close examination, viewing the flowers requires a telephoto lense and remains a challenge even then. Anderson has concluded on the basis of molecular studies carried out by Wallace that both *Jasminocereus* and the low-lying *Brachycereus*

Figure 2.239. Fruit of Jasminocereus thouarsii. *The husk is tougher than on any other columnar I have seen.*

nesioticus (a lava-loving, clustering Galápagos endemic) are derived from *Armatocereus,* the mainland sausage cactus growing closest to the Galápagos.[150] *Jasminocereus* exhibits a clear phenotypical resemblance to *Armatocereus carwrightianus,* though it is less spreading, but *Brachycereus* shows no resemblance at all. (Fig. 2.239.)

Azureocereus and *Browningia*

Azureocereus is a much smaller genus, with only two members, which probably represent the same species. Recent cactologists have referred both species to Browningia. I stick stubbornly to *Azureocereus,* only because the blue color of younger plants sets it off from any *Browningias* I have seen. (Fig. 2.240.)

Browningia is a strictly Andean, mostly Peruvian genus of at least eight species. Almost all grow in Peru, except for the poorly known *B. caineana* (Cárdenas) D. R. Hunt of lowland Bolivia and, perhaps, northwestern Paraguay. *B. candelaris* is perhaps the best known, a strange tree from the Atacama Desert of Chile and the driest portions of southern Peru. Most *Browningias* are columnar, the tallest reaching 10 m in height.

Azurocereus

Azureocereus hertlingianus (Backeberg) Backeberg
saguaro, *sanquey* (Quechua)

Figure 2.240. Azureocereus hertlingianus *showing blue color and rib configuration.*

How this unusual and quite restricted cactus came to have the same common name as the giant of the Sonoran Desert is a mystery. It is universally called that in the environs of Ayacucho, Peru, where it is well known and commonly planted as an ornamental. Large populations grow near the indigenous city of Huanta and on steep desert hills near the Quechua village of Simpapata, about one hour's drive from Ayacucho at around 2450 m elevation. The population there appears to be robust, with a healthy mix of old and young plants. The older saguaros are densely armed with long, thick spines, which seems a pity because the spiny armor makes the cacti look a bit ratty and obscures the bluish tint that renders the juveniles so handsome. The plants reach about 7 m in height at Simpapata, somewhat taller near Huanta. (Fig. 2.241.)

In addition to its startling bluish color, *Azureocereus* demonstrates two oddities: first, its fruits are black, the only example of that color of fruit among columnar cacti I know, and, second, the areoles occupy a depression on the surface of the ribs—that is, they are concave, not raised, as is usually the case in columnar cacti.

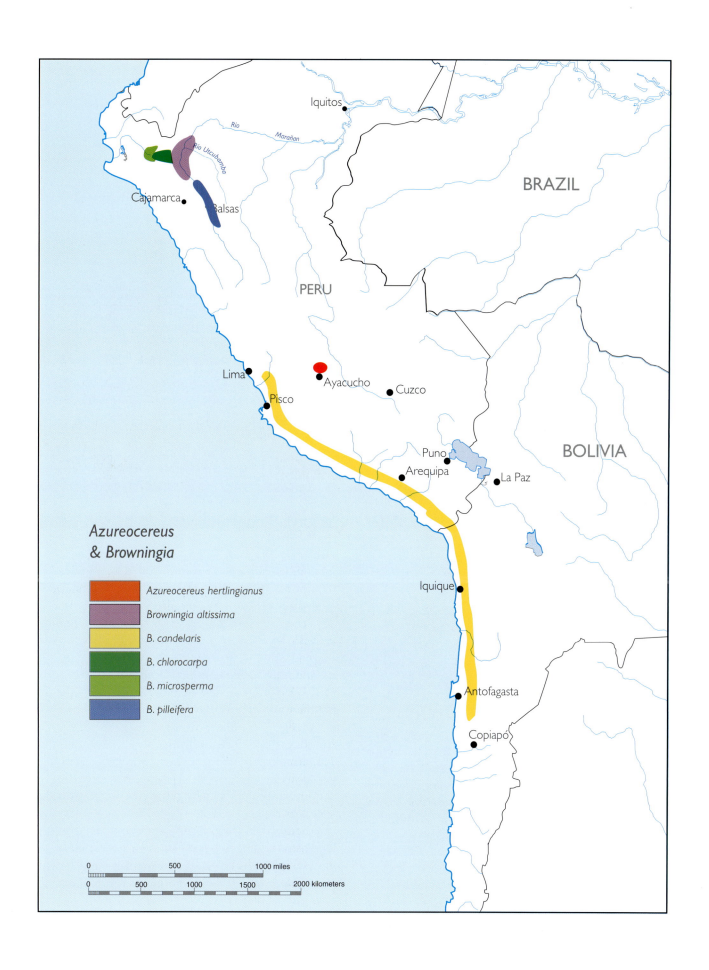

Iquitos

Río

Rio Utcubamba

Río Marañan

BRAZIL

Cajamarca

Balsas

PERU

Lima

Pisco

Ayacucho

Cuzco

Puno

Arequipa

BOLIVIA

La Paz

Azureocereus
& Browningia

Iquique

Antofagasta

Copiapó

Azureocereus hertlingianus

Browningia altissima

B. candelaris

B. chlorocarpa

B. microsperma

B. pilleifera

| 0 | | 500 | | 1000 miles |

| 0 | 500 | 1000 | 1500 | 2000 kilometers |

Figure 2.241. Azureocereus hertlingianus *near Simpapata, Peru.*

Figure 2.242. *Toy cannon sold in Ayacucho, Peru, market, made from* Azureocereus hertlingianus *wood.*

Figure 2.243. *Unripe* Azureocereus hertlingianus *fruit.*

These saguaros possess an unusually hard wood. Artisans in the region cut the dead branches into sections and fashion toys and miniatures that are marketed in Ayacucho and beyond. The fruits are said to be sweet and good for sore kidneys. (Figs. 2.242 and 2.243.)

Browningia

Browningia altissima (F. Ritter) Buxbaum

In Amazonas province of northern Peru, where the great Río Marañon veers north and slightly east from its northwestward course, the thornscrub desert receives rainfall of less than 500 mm, producing an arid climate close to the equator. This microclimate extends southeastward along the Marañon for a few hundred kilometers, westward for about 160 km along the Río Chamayo, a Marañon tributary, and southward into the lower Río Utcubamba valley, where *B. altissima* mingles spectacularly with *Espostoa utcubambensis*. In this small desert, columnar cacti have proliferated dramatically, rivaling the Valle de Tehuacán in Mexico for diversity of cactus species. The tallest of these arid species is *Browningia altissima*, a very common (in its habitat), spreading columnar that reaches nearly 10 m in height, with dozens of branches. (Fig. 2.244.)

Although *B. altissima* is not as bald as the upper branches of *B. candelaris* and the older branches of *B. pilleifera*, it is semibald. Most of the areoles of mature plants lack spines, and those that retain them have but one long central, drooping spine (3–4 cm), guarding the plant rather like a low-paid security guard, sometimes along with a couple of additional radial spines. The areoles on the younger branches are nearly flush with the bump on which they are located. With age, they decrease in area and appear to be recessed into the branches. (Another reason why *Azureocereus* and *Browningia* should probably be grouped into the same genus!) Hundreds of these great plants combine to effect a design that decorates the landscape with a most pleasant and characteristic appearance. *B. altissima* can be seen growing up steep hillsides well south of Bagua in the Cañon Utcubamba.

Browningia candelaris (Meyen) Britton & Rose
tinaja, soberbio, candelabro

B. candelaris is a superb example of an Atacama Desert plant and its environs, a desert as dry as any in the world. This mostly rainless north-south strip of northern Chile and southern Peru, 3700 km long, derives its aridity from a combination of two desert-inducing phenomena: a cold-air climate produced by the chilly offshore Humboldt Current, which flows northward and extinguishes any and all rain-generating convection, and a rain shadow desert produced by the towering Andes, which cut off all or nearly all Atlantic moisture from arriving at the west slopes of the Cordillera.[151] Coastal precipitation is less than 5 mm per year (that's less than one-fifth of an inch), and higher elevations may see years without measurable rainfall. In coastal northern Chile, rain never falls. As noted earlier, I once asked a taxi driver in Antofagasta, Chile, the date of the last known rain. He replied that it was several million years earlier.

Figure 2.244. Browningia altissima, *Río Chamayo, Peru.*

This most odd, sometimes tall cactus (up to more than 7 m, but usually much shorter) appears to tolerate the absence of rain like no other large plant. It grows only between just under 1600 m and 2800 m elevation in northern Chile and southern Peru nearly as far north as Lima. The lesser altitude appears to represent the lowest elevation where the excruciatingly rare summer rains reach. Years may pass between rains, but even in their absence the plants continue to grow, though they may become scraggly and appear to be dead after years of rainless days. Where and when rains do occur, the plants turn green and appear robust and attractive. The range of *B. candelaris* is more than 1300 km north to south, even greater than that of the American saguaro, but the habitats it traverses in that range are far less varied than the habitats crossed

by *Carnegiea.* In Cañon Tinajas near Lima, the plants appear to tolerate up to 200 mm of rainfall and share the landscape with numerous other plant species, including the cacti *Armatocereus matucanensis, Espostoa melanostele, Haageocereus pseudomelanostele,* and *Melocactus peruvianus.*[152] On the very dry hills near Arequipa, *Browningias* are found in association with *Weberbauerocereus weberbaueri, Neoraimondia arequipensis,* and *Arequipa* sp. In other places, they appear at times to be the only plant growing in their habitats. They seem to thrive on ridgelines and are often visible from miles away, resembling aliens outlined against the skyline, the only identifiable objects crowning the sere hills. (Figs. 2.245 and 2.246.)

The adult cacti are so strange that they are nearly impossible to confuse with other species. In some areas, but not

Figure 2.245. Upper branches of Browningia candelaris, *Cañon Tinajas, Lima, Peru.*

all, juvenile growth is heavily armored with stiff, stout spines—among the meanest anywhere—growing in a dense network. The adult growth at above 2 m or so is unarmed, however, so the plants appear as though the top of a gnarled cactus plant with arms growing every which way had been screwed onto the straight trunk of an entirely different species. The trunk seems well suited to discourage predation by megaherbivores and even by smaller ones. The buds and fruits are found only on the spineless adult branches, high enough to escape the grazing even of guanacos and extinct camels. Spines continue to grow on the trunk throughout the life of the cactus. The fruits appear to be covered with scaly leaves, most unusual for cactus fruits and especially on a plant that seems perfectly adapted to withstand many months of no rain.

The soberbio's rate of growth may be microscopic. Graham Charles cites a specimen in his collection (in a greenhouse in England) that reached a height of only 25 cm after 25 years.[153] The slow-growing saguaro (*Carnegiea gigantea*) is a shooting star by comparison, for a saguaro of such age would typically measure a meter or more in height. A 4.5-m-tall soberbio would be 450 years old at such a rate of growth. I suspect that plants grow even more slowly and reach a much greater age in parts of the their range, where rains are as rare as leap years. In more moist regions, such as upper Cañon Tinajas, young plants appear

Figure 2.246. Browningia candelaris, *Cañon Tinajas, Lima, Peru.*

to grow more rapidly and bear a startling resemblance to slender young saguaros. (Fig. 2.247.)

The soberbio fruits are sweet and widely gathered, weighing up to 200 g, or half a pound. Fruit production is spotty from year to year, however, and harvests are erratic, so the wild plants are not usually considered a reliable source of food. *Browningias* appear to be especially vulnerable to damage by livestock grazing. In spite of the desirability of their fruits, populations of soberbios do not appear to be reproducing well, for in parts of Peru all or nearly all young plants are eaten or trampled by goats. The rate of seedling recruitment is difficult to ascertain because the conditions for successful germination of seeds appear to occur very rarely, and seedlings are gobbled up almost as soon as they appear. Young plants, those less than 2 m tall, are rare indeed.

Carlos Ostolaza believes that *B. candelaris* played a major role in pre-Columbian coastal civilizations. Based on his research, he hypothesizes that the sensational geoglyph called the Candelabro of Paracas, excavated in the desiccated soil above the ocean near this Peruvian coastal city, is a representation of the cactus.[154] Its age and purpose are the object of considerable discussion. (Fig. 2.248.)

The wood of the soberbio trunk is fairly strong and light, ideal for construction. This, too, does not bode well for the cactus, for many thousands of plants with the misfortune to grow near human settlements have been sacrificed for their lumber, particularly around the coastal mines of northern Chile. In addition, the woody trunks burn well and are the only source of firewood over hundreds of square kilometers. Near the city of Arequipa, squatters' lots creep up the steep hillsides, and the trunks from candelabras make fine temporary posts for shacks that establish the squatters' legitimacy. The plant's slow-growing nature makes it likely that in habitats within a few kilometers of human settlement, the cactus will gradually disappear as those individuals that survive anthropogenic onslaughts die and no young plants survive to replace them. (Fig. 2.249.)

Browningia chlorocarpa (Kunth) W. T. Marshall

While driving east of Abra Porculla northeast of the coastal city of Chiclayo, Peru, I spied these large cacti growing on a steep hillside. The landscape east of the pass is very dry desert, unlike the more humid regime west of the pass. In fact, the higher one gets on the Río Chamayo, the Marañon tributary, the drier it becomes, giving rise to several species not known from the semiarid lands to the west of the pass. Although the plants were numerous, I was unable to identify them, for they bore no flowers or fruits and could have passed for a *Cleistocactus* (a friend of mine incorrectly

Figure 2.247. Browningia columnaris, *Lauca Valley, Chile. Note the absence of other vegetation.*

identified it as such.) I called it *Cleistocactus roezlii* in a published article.[155] Graham Charles saw the photograph in the article and notified me that the cactus was undoubtedly *B. chlorocarpa* instead. Apart from the photograph, I have not garnered much information about this handsome plant. Numerous specimens were to be seen in the same locale at around 1500 m elevation, but no person was in the vicinity to provide me with any details. I will just have to go back someday. (Fig. 2.250.)

Browningia microsperma (Werdermann & Backeberg) W. T. Marshall
faica

Among the tallest of Andean cacti, these gracefully erect plants reach 10 m in height. It is difficult to determine

Figure 2.248. The Candelabro, geoglyphic figure on the Paracas Peninsula, thought by some to represent Browningia *candelaris.*

their natural color because they appear to prefer the moist western slopes of the coastal Peruvian Andes, where they are host to a wide variety of lichens, mosses, bromeliads, and, I assume, orchids. I found many of them growing some 10 km west and below Abra Porculla between Bagua and Chiclayo, where numerous large plants have survived the clearing. Their preference for steep slopes has probably saved them up to this point, but out of desperation poor farmers are planting on steeper and steeper slopes. If the cactus fruits are edible, it must be a chore to gather them and perhaps impossible, for they grow at or near the tips of the branches, and the giant cacti themselves grow only on steep, rocky hillsides. I suspect that *B. microsperma* is truly in danger of extinction, for its ecological requirements are very narrow, and even the steepest slopes in its range are slowly being cleared for planting. Although the plants are common in a band at more than 1000 m elevation (roughly 1400–1500 m), their habitat is under siege. (Fig. 2.251.)

Browningia pilleifera (F. Ritter) Hutchison
yonco

On the bajadas and slopes above the hot, somnolent village of Balsas on the Río Marañon appear thousands of handsome, nearly bald, goblet-shaped columnars. Seldom does the rational cactologist have an urge to stroke a cactus, but the branches of this fine columnar appear so smooth that I found myself petting it in wonder. The cuticle is more like

Figure 2.249. Browningia *trunks used as corners for a squatter's hut, Tiabaya, Arequipa, Peru.*

Figure 2.250. Browningia chlorocarpa *east of Abra Porculla, northern Peru.*

that of a watermelon than that of a cactus, even though it usually bears the scars of insect attacks. The areoles are evident on new growth within 20 cm or so of the apex, in the same vicinity as the fruits, but the felty excrescences, such telltale indicators in most cacti, gradually disappear, leaving only a scar behind. Whether this happens because the felty growth simply disintegrates with time or because the cactus jettisons it, I cannot say. The plant's overall appearance is unusually clean and friendly. Taxonomic descriptions of the species mention spines, but those individuals near Balsas mostly lack spines. The color of individual plants is unusual, ranging from a grayish green

to a pale brownish green that could be ugly in a lesser plant, but in the case of the *yonco* is handsome, indeed. (Fig. 2.252.)

The fruits are edible, strikingly similar to those of *Escontria chiotilla* of southern Mexico both in shape and in that large scales instead of spines protect the husks. (This is also true of the fruits of *Azureocereus*.) Few people in Balsas consume the fruits, and the fruits were not ripe during my visit, so I had no opportunity to test them. The wood, though brittle, is extraordinarily hard, resisting attacks by machete.[156] The cut surface of the cuticle turns black, as is the case with the genus *Pachycereus*. (Fig. 2.253.)

Figure 2.251. Browningia microsperma *near Abra Porculla, northern Peru. Frequent fog allows numerous lichens and bromeliads to grow on the plant.*

Figure 2.252. Browningia pilleifera *near Balsas, Río Marañon, Peru.*

Figure 2.253. Fruits of Browningia pilleifera, *Balsas, Rio Marañon, Peru.*

Figure 2.254. Cereus hildmannianus, *Tucson, Arizona.*

The *Cerei*

The genus *Cereus* has proliferated over much of tropical South America east of the Andes, sometimes poking through to the western slopes. It includes two of the largest and tallest South American columnar cacti, *C. lamprospermus* of eastern Bolivia and *C. hexagonus* of the northern lowlands of the continent. Species in the genus also produce flowers that are among the gaudiest in the cactus family. I have selected four representative species for this book.

Cereus hildmannianus K. Schumann
[= *Cereus peruvianus* (L.) J. S. Muell.]
[= *Cereus uruguayanus* R. Kiesling]
Peruvian apple cactus

This tall, spreading cactus is well known to cactus fanciers throughout the world. It is ridiculously easy to propagate, grows quickly (a young plant will easily grow more than a foot per year), tolerates moderate frost, and produces numerous showy white flowers 15 cm across. The goose-egg-size fruits contain a delicately sweet white pulp with delightfully crunchy black seeds. They possess the added virtue that their husks are free of spines, so one can simply grab one and twist it off the branch without worrying about impalement. The plant's rapid growth and fruit production apparently have made it attractive throughout the neotropics. It is so commonly grown that its origins are unclear and in dispute. Witness, for example, the common name and the former species names. It might have originated in Peru or in Uruguay or in neither.

Even more intriguing, the apple cactus is widely touted as having the power to correct physical ailments caused by electromagnetic radiation. How and why this particular virtue was discovered are unclear, but the plant is widely sold potted with the recommendation that it be located near a computer screen or television in order to reestablish an electromagnetic equilibrium upset by the device. Or some such thing.

The apple cactus also grows readily from cuttings, sprouting quickly and growing rapidly into a tall columnar. I planted a cutting 30 cm tall in my front yard in 1990. By the year 2000, it was 6 m tall, with more than a dozen arms. Even more important, it often produces more than one hundred delectable fruits each year, all of which act as bird magnets. I have watched hummingbirds nibble on the fruit after larger birds (especially Curve-billed Thrashers and Gila Woodpeckers) have made a hole in the husk. I must monitor the ripening fruits vigilantly, for if I delay harvesting them by so much as one day, the birds will attack them and leave me with only half the delectable pulp. Once harvested, the ripe fruits must be eaten within twenty-four hours, or they begin to ferment. They last longer if harvested prior to ripening, but the quality appears to be damaged by premature collecting. I have never tried drying the fruits or making preserves or wine, but I encourage others to do so. (Fig. 2.254.)

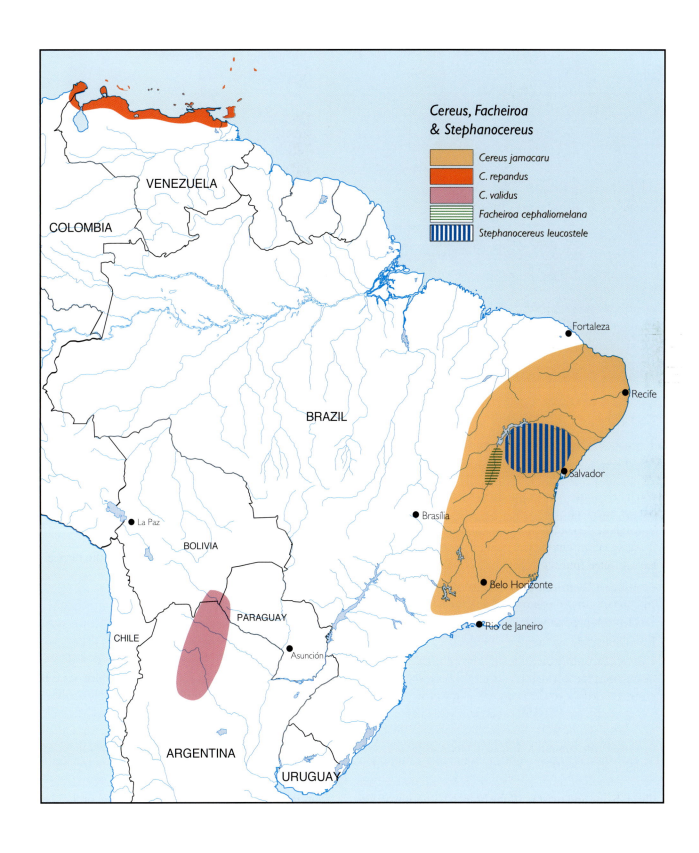

Cereus, Facheiroa
& Stephanocereus

- Cereus jamacaru
- C. repandus
- C. validus
- Facheiroa cephaliomelana
- Stephanocereus leucostele

COLOMBIA

VENEZUELA

BRAZIL

Fortaleza

Recife

BOLIVIA

La Paz

Brasília

Salvador

Belo Horizonte

Rio de Janeiro

PARAGUAY

CHILE

Asunción

ARGENTINA

URUGUAY

Figure 2.255. Mandacarú (Cereus jamacaru), *Minas Gerais, Brazil.*

The flowers are equally appealing to winged creatures of the night; I have watched the 15-cm-wide flowers open, usually after ten o'clock. All manner of pollinators are attracted, including hummingbird moths at night and carpenter bees, honeybees, houseflies, wasps, and a host of lesser arthropods in the daytime, but I believe nectar-eating (*Leptonycteris*) bats are the principal pollinators. They are difficult to photograph without specialized equipment, however, for they dart in and out of the flowers rapidly, lingering less than a second to drink the nectar.

The plants are dramatically responsive to warm-weather rain. Following any substantial summer (or warm-weather) rain, they will produce buds that seem to sprout as if by magic from the areoles of the ribs. The buds grow quickly, shooting out from the branches like so many small snakes. Within two weeks, the flowers open, always at night. By mid-morning, all have closed, so only early risers are rewarded by the sight of dozens of flowers simultaneously opened. The fruits usually ripen within a month. Buds may erupt well into fall, even early winter, if sufficient rains fall. If the fruits do not receive adequate moisture, however, they will remain small, and the pulp will be less sweet and tasty than the fruits of plants adequately irrigated. During dry times, the buds often fall off prior to opening, thus sparing the tree some expenditure of energy.

Experimental growing of the plants in the harshly dry Negev Desert of Israel has proved promising for fruit production.[157] Plants irrigated with freshwater produced more than 20 kg (44 lbs.) of fruits only six years after planted as cuttings, which compares favorably with dryland fruits of more temperate species.

Cereus jamacaru A. P. de Candolle
mandacarú, surucucú

The best-known cactus of Brazil, this giant is common throughout the eastern portion of the country. It ranges from a scraggly 3-m-tall single trunk in the driest sertão to a huge, spreading tree in more well-watered regions. It commonly exceeds 10 m in height, reportedly reaching a colossal 20 m, though I am sceptical of such reports. Its fruits may bear red pulp and are relished everywhere, though they must compete with the cornucopia of other fruits in this fruit-rich country. In some areas, the wood of the trunk is said to be dense enough for lumber, though reports on that score are inconsistent. A farmer with a massive plant growing in his yard explained that though he and his family eat the fruits, they value the tree the most because when it flowers (in October and November): it means that rain will soon fall, and they will be happy. (Fig. 2.255.)

The mandacarú is widely planted horticulturally, used as an ornamental in city parks and medians, and grown in patios. Although wild individuals appear to be common throughout the sertão, especially so in the more moist *caatinga* (the scrubby, thorny vegetation of the sertão), it is difficult to distinguish between wild and cultivated plants. (Fig. 2.256.)

Cereus repandus (Linnaeus) P. Miller
kadushi

It seems odd that a large cactus would be the most important native plant on Caribbean islands, but the kadushi is the outstanding tree on the ABC Islands of the Netherlands Antilles. Kadushis grow into huge plants—up to 12 m tall, with hundreds of branches. They have clearly defined, strong trunks and might invite climbing were it not for their myriad long and potent spines. Unlike many of the tribe Pachycereeae, trunks of the tribe Cereeae tend to retain their spines well into adulthood. The spines are rigid and well developed. (Fig. 2.257.)

Kadushis appear to find the slopes and hillsides of Bonaire especially to their liking. They grow there in forests that rival the pitayales of Sonora and the tetecheras of Puebla and Oaxaca in density. They also are successful in recruitment. It is not unusual to see overgrazed soils overgrown with hundreds of kadushi seedlings forming an impenetrable thicket.

Kadushis, like so many columnars, are sources of fresh fruit and lumber. Prior to the globalization of lumber, many beams and door and window frames were fashioned from their strong trunks. The trunks were the only source of straight lumber on the islands, so the wood was coveted and widely exploited. The wood can be cut into planks and, though tough, can be planed and sanded to a fine finish. The few older buildings on Bonaire and Aruba that remain as testimony to pre–World War II society often sport lumber from the kadushi. Fortunately, the islands (especially Bonaire) still support huge numbers of the large plants, and the switch from native to imported lumber for constructing buildings was a godsend to the kadushi population.

Since the early days of sugar plantations (incorporating slave labor) in the Netherlands Antilles, kadushi cuttings have been planted into living fences. Many of these fences are still to be seen, now composed of many-branched plants up to 10 m tall. However, the yato *(Stenocereus griseus)* appears to grow more uniformly and straighter and has thus become more common than the kadushi for fencing on these islands. Yato living fences appear to be a much more recent phenomenon, so the switch from kadushi to yato may not have happened in prehistory. Yato fruits are

Figure 2.256. Mandacarús in median, Itaobim, Minas Gerais, Brazil.

sweet but spiny and must be harvested gingerly, and the lumber is inferior to that of the kadushi, so fence builders needed to consider these fences' long-term yields, a consideration the world's fence builders have not commonly contemplated.

The purple-husked kadushi fruits appear all year round, but are associated primarily with the rainy months of December and January. They are small compared with those of apple cacti—usually no more than 6 cm long and 3 cm wide. With some supplemental watering, the fruits may increase in size. The pulp is white with black seeds and has a lightly piquant flavor combined with a somewhat bland sweetness. The fruits have no spines, so they are simply pried from the branches and peeled. I have seen no evidence

Figure 2.257. Kadushi (Cereus repandus), Washington-Slagaabi National Park, Bonaire.

of selection of plants for their fruits, though this process could well have occurred in other locations to which the plants were transplanted. (Fig. 2.258.)

The most unusual use of the kadushi, however, is its role as a vegetable. A heavy soup with kadushi as a base was until recently nearly a staple in the islanders' diets. A Bonairan woman described to me the process of cooking it. The spines are removed from the branches (a laborious task). The green exterior flesh is then carved from the branches, collected, and separated from the inner, more pithy tissue, which is boiled, then simmered with pork, snails, and fish, and seasoned. The resulting soup is still considered a delicacy. Few women continue to cook it, however, probably because of the heavy labor requirement. Others have reported that the plant tissue can be dried and ground into a powder that can be stored indefinitely.[158]

Cereus validus Hayworth
ucle

Associated with the unquillo *(Stetsonia coryne)* and the cardón *(Trichocereus terscheckii)* in the Chaco forest of South America is the *ucle, Cereus validus.* It bears a close resemblance to the genus *Armatocereus* of Ecuador and Peru, its arms forming numerous sections tapered like

Figure 2.258. Kadushi fruits, Aruba.

Figure 2.259. Ucle (Cereus validus), *montane forest, Salta, Argentina.*

sausages (see fig. 2.231). These rather thin, frail-looking, but attractive columnars reach a height of 6 m, perhaps higher under favorable circumstances. At times, they appear to be swallowed up by the surrounding vegetation, but are tolerant of partial shade and still seem to flourish from the semiscrub of the Chaco to the moist montane forests at 1400 m.

Ucles are well known for their reddish fruits relished by bird and human alike. The plants flower in spring, and the fruits ripen in summertime. (Fig. 2.259.)

Pilosocereus and Other Bearded Cacti of Brazil

Pilosocereus is by far the most wide-ranging genus of columnar cactus, with species flourishing throughout the tropics, south to the Tropic of Capricorn[159] and well north of the Tropic of Cancer. It is also the largest columnar genus, with perhaps thirty species in at least seven differ-ent countries and on numerous islands in the Caribbean. Brazil is a hot spot of *Pilosocereus* evolution. Just how many species it hosts is subject to considerable debate between the experts. For this book, I have chosen the taxonomy of Brazilian cactologist Daniela Zappi.[160]

In spite of the remarkable proliferation of *Pilosocereus* and related species of bearded columnars *(Coleocepha-locereus, Facheiroa, Microanthocereus, Siccobaccatus)* in Brazil, their ethnobotanical uses are limited. The strange, but legitimate (though skinny) columnar *Stephanocereus leucostele* also makes spotty appearances, but I have been unable to gather any ethnobotanical information about it. Although highly local uses may be made of these often bizarre-appearing plants, they do not seem to be of great importance in contemporary or historic Brazilian cultures. There are practical reasons for this nonuse: many, if not most, of the species are restricted to small or even tiny distributions, often on substrates such as the contorted pre-Cambrian limestone formations called *bambui* that are inaccessible to livestock and uncomfortable for humans.

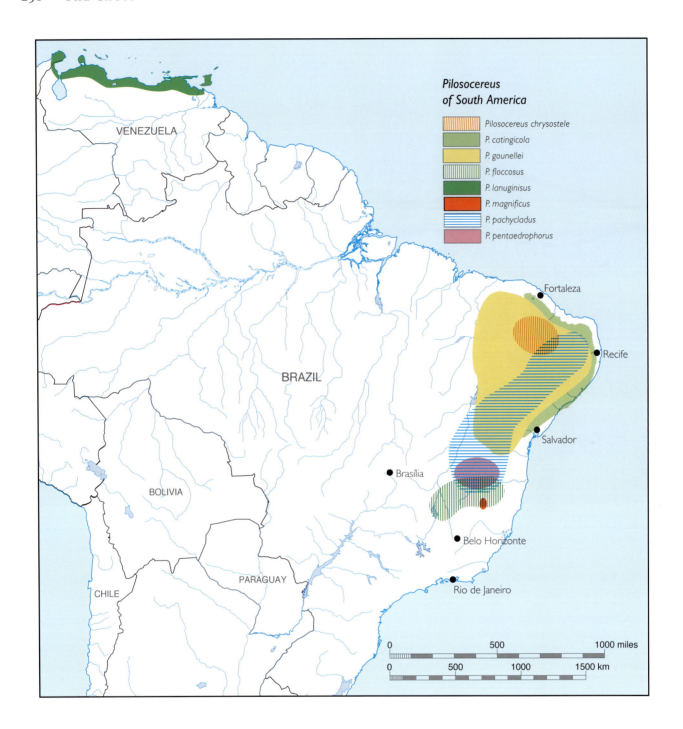

The species have evolved into tiny niches, each perhaps with its own microenvironment, often only a few kilometers in radius. In some cases, only single populations on solitary rock outcroppings are known to exist. Apart from the cosmopolitan mandacarú (*Cereus jamacaru*), Brazil has no broad-ranging columnar cactus.

The same disclaimers apply to my disappointingly brief presentation of Brazilian columnar cacti. Although tall cacti are well represented in the nation, the distribution of each species tends to be very limited. Some are found only on highly specialized substrate in extremely remote locations, requiring journeys of hundreds of kilometers. Visiting and photographing all the species would require a huge allocation of time and resources. In several cases, a species may be limited to a few individuals on a torturously inaccessible outcrop kilometers from the nearest town and scores of kilometers from the nearest airport. One needs to be a Brazilian like Zappi to accomplish an

Figure 2.260. Facheiroa cephaliomelana *in fruit, growing on bambui, a pre-Cambrian limestone, Bom Jesus da Lapa, Bahia, Brazil.*

adequate cataloging of Brazilian columnars. Consequently, as a foreigner, I have selected a cross section of the genus as representative of Brazil's columnar cacti.

Brazilians use columnars, but sparsely. All species appear to bear fruits that are easily edible and are horticulturally desirable as well. For example, although I could gather no

information whatever about *Facheiroa cephaliomelana* growing out of inhospitable bambui near Bom Jesus da Lapa, Bahia, I managed to pluck a fruit and found it to be subtly sweet and satisfying. (Fig. 2.260.)

Brazilians use the common name *facheiro* to refer to columnar cacti.

Figure 2.261. Pilosocereus chrysostele, *Ceará, Brazil.*

Figure 2.262. Pilosocereus chrysostele *growing from old* fazenda, *Ceará, Brazil.*

Pilosocereus chrysostele (Vaupel) Byles & G. D. Rowley
cardeiro; rabo-de-raposa

Within its range, which is in the arid sertão, the interior of northeastern Brazil, this plant grows abundantly on granite pediments and hillsides. It is a noticeable part of the vegetation in portions of central Ceará, where some of its many branches reach up to 6 m in height. It usually lacks a well-defined trunk. The common name *cardeiro* is the equivalent of the Spanish *cardón*, derived from the name for thistle. Probably because this cactus is common near villages in the sertão of Ceará, *sertãoeiros* attribute several important uses to it. For sore throat, a paste is made from the flesh and applied to the outside of the neck or mixed with hot water for a tea. For cleansing the liver and as a remedy for diabetes, the spines are sliced off a branch, then pieces of center flesh are boiled for a while, and the resulting decoction is drunk as a tea. In times of severe drought (moderate droughts are common in the sertão), ranchers may slice or burn off the spines and lop off branches for livestock food. (Fig. 2.261.)

These thin but handsome plants recruit well and colonize odd places, including abandoned and decaying buildings. (Fig. 2.262.)

Pilosocereus catingicola (Gürke) Byles & G. D. Rowley
facheiro

Large and handsome plants of this species appear surprisingly on coastal dunes along vast stretches of beach in the states of Ceará and Sergipe, as well as in other habitats in northeastern Brazil. Though the plants growing on dunes seldom exceed 6 m in height, elsewhere they become massive, reaching 10 m, with a woody trunk that may serve as lumber. Although the fruits are edible, I was not able to obtain any ethnobotanical information about them. (Figs. 2.263 and 2.264.)

Pilosocereus floccosus Byles & Rowely

Restricted to gneiss or ancient granite outcroppings in the province of Minas Gerais, *P. floccosus* has a highly restricted range, a few favorable locations there. In

Figure 2.263. Pilosocereus catingicola *and the author. Sand dunes, Atlantic Ocean, Ceará, Brazil.*

these limited habitats, only small numbers of the cacti are to be found. On occasion, a plant will grow into a paradigmatically columnar form, reaching at least 7 m in height. This cactus is notable for its handsome green color and the bright reddish pulp of its fruits. It has the added virtue of sharing its open granite habitats with the startling *Melocactus bahiensis.* (Figs. 2.265a, 2.265b, and 2.266.)

Pilosocereus gounellei (F. A. C. Weber) Byles & Rowley
xique-xique

This cactus is common in the northern sertão, where it resembles the chollas of the Sonoran Desert (especially *Cylindropuntia fulgida*) in responding well to overgrazing and seizing additional acreage for itself. It barely reaches 2 m tall in the very arid sertão of Ceará, but becomes considerably taller elsewhere, a bona fide columnar. In times of severe drought, ranchers burn off or cut off the vicious spines and feed the branches to livestock. In scrawnier plants, the cottony tufts usually diagnostic of *Pilosocereus* are absent. (Fig. 2.267.)

Pilosocereus magnificus (Buining & Brederoo) F. Ritter

It seems a shame that a plant of such agreeable appearance should be one of the columnar cacti most limited in distribution. *P. magnificus* appears to grow only in a couple of granitic outcroppings along the Río Jequitinhonha in northern Minas Gerais state, where it seems to be flourishing. The young cacti apparently prosper under nurse plants, which may be scarce on the solid granite substrate these cacti seem to prefer. The outcrops are few in number, so there are few sites where these cacti can be found. The branches are bluish and emerge from the base. I was unable to determine any common name. (Fig. 2.268.)

Figure 2.264. Areoles of Pilosocereus catingicola *showing development of hairs.*

Figure 2.265a. Pilosocereus floccosus, *Itaobim, Minas Gerais, Brazil. An unusually upright and columnar form, perhaps responding to dense vegetation around it.*

Figure 2.265b. Pilosocereus floccosus *in fruit, near Col. Murta, Minas Gerais, Brazil.*

Figure 2.266. Melocactus *cf.* salvadorensis *growing in conjunction with* Pilosocereus floccosus *on granite outcrops, Minas Gerais, Brazil.*

Figure 2.267. Pilosocereus gounellei, *as common in parts of the sertão as chollas are in the Sonoran Desert, achieving a height of more than 2 m (7 ft.).*

Figure 2.268. Pilosocereus magnificus, *near Itinga, Minas Gerais, Brazil. It has the smallest distribution of any columnar cactus described in this volume, being confined to one or two granitic outcroppings along the Río Jequitinhonha.*

Figure 2.269. Pilosocereus pachycladus. *This tall, skinny, nearly beardless cactus appears on the scene quite suddenly in south-central Bahia, Brazil, but is nowhere common.*

Pilosocereus pachycladus F. Ritter
facheiro azul

Even though this cactus may reach 10 m in height, it is possible to overlook shorter plants due to their dainty or slim profile. A light blue-green color, it prefers the dense scrub of well-developed caatinga, where it shares habitat with *Cereus jamacaru* and an occasional *Stephanocereus leucostele*. It often develops a distinct trunk and may grow dozens of branches. (Fig. 2.269.)

Pilosocereus pentaedrophorus (Cels) Byles & Rowley
facheiro

I found only one population of this handsome cactus. It consists of a dozen or so mature plants and many more younger ones on the summit of a granitic outcropping in

Figure 2.270. Pilosocereus pentaedrophorus, *southern Bahia, Brazil. It grows only on outcroppings of ancient granite and in a very confined area.*

the midst of the vast sertão of southern Bahia province. I noted the plants' silhouettes from a distance, but was forced to fight my way through scrubby, tick-infested thornscrub to reach them. Alas, for miles around I looked in vain for some soul to provide me with any local knowledge of the plants, but none was to be found. Zappi notes several other locations where it can be found, but all appear to be quite limited in area.[161] (Fig. 2.270.)

Pilosocereus tweedyanus (Britton & Rose) Byles and Rowley

cactus

This Ecuadorian species carries the improbable name of a botanist. It is prominent on hillsides near the central coast of Ecuador and in drier interior valleys. It appears to favor tropical deciduous forest in the southwest, competing well

with other trees in that habitat. Indeed, it appears to be the dominant plant on slopes near the coast and inland for several kilometers. It is difficult at first glance to pronounce the plant a cactus, for many hundreds of these 7–10 m tall, many-branched columnars become nearly covered with vines and creepers. The primary culprit is a morning glory vine (*Ipomoea* sp.) that grows rapidly throughout the coastal lowlands with the onset of the rainy season. Whether this climber is expanding and will have a deleterious effect on the cactus will soon become evident. During the dry or dormant season, the vines lose their leaves, and the numerous cactus-vine complexes appear to be amorphous, undifferentiated gray-brown masses jutting up from the green hillsides. Usually, though, white caps visible from afar identify the cacti. Numerous long, entwined white hairs on the branches produce a pseudocephalium responsible for

Figure 2.271. Pilosocereus tweedyanus, *western Ecuador. Many thousands of these plants grow in the coastal lowlands of southwestern Ecuador.*

this effect. Plants unencumbered by vines tend to develop a spreading habit in candelabra fashion. Occasional individuals have a single trunk up to 3 m tall from which arms then divagate. The *cactus,* the drab name by which the plant is known in the region, produces an edible pinkish fruit the size of a golf ball. The fruits are embedded within the woolly white pseudocephalium on the cactus and are detectable as ripe only by the flash of pink visible through the white. They can be easily harvested by prodding them with a pole. When I tasted the sweet fruits, it seemed odd to me that natives in general denied that the plant was good for anything. One fellow explained that the spines interfered with eating the fruit, but when I demonstrated the nearly complete lack of spines, he acknowledged that the problem wasn't that at all. I picked a fruit and ate it, however, and showed the empty husk to a couple of peas-

ants. They acknowledged that the fruits were good and that, yes, they did eat them. In the abundance of local tropical fruits, however, the *P. tweedyanus* fruits are probably too difficult to harvest and deemed not worth the effort. In arid lands, where humanly useful plants are scarcer and gleaning food from the land is more difficult, cacti play a more significant role, and natives distinguish more clearly among species. (Figs. 2.271 and 2.272.)

Facheiroa cephaliomelana Buining & Brederoo

These odd plants barely qualify for columnarhood. However, they at times exceed 2 m in height and resemble columns, so I have included them. Plants of this genus tend to be highly localized and grow only on pre-Cambrian limestones known in Brazil as bambui. *F. squamosa* reaches 8 m in height. The examples in figure 2.273 grow on a small mountain of bambui, the only feature in the broad, flat valley of the Río San Francisco in southwest-central Bahia. Their fruits are edible (I will attest), but they do not appear to be used, even in the town of Bom Jesus da Lapa, which surrounds the inselberg. (Fig. 2.273.)

Stephanocereus leucostele (Gürke) A. Berger

Its strange appearance alone qualified the *Stephanocereus* for inclusion in this book. I have no ethnobotanical information about it, not even a common name. It is a bit of a loner, usually a single stalk with tightly appressed branches, scattered widely in the caatinga of southern Bahia. The plants tend to appear whitish, and only occasionally does the green cuticle of the trunk show through the heavy coat of bristles and hairs. At intervals of somewhere around a half meter, shorter toward the apex, it sports a ringlike cephalium or pseudocephalium that makes it appear as though it was assembled in sections, rather like *Equisetum*. Other Brazilian semicolumnar cacti of the genus *Arrojadoa* display similar rings. (Fig. 2.274.)

Figure 2.272. Fruit of the Pilosocereus tweedyanus, *Ecuador.*

Figure 2.273. Facheiroa cephaliomelana, *Bom Jesus da Lapa, Bahia, Brazil.*

Figure 2.274. Stephanocereus leucostele, *southern Bahia, Brazil. The rings on this strange plant may be correlated with its age. It grows only in an area of a few hundred square kilometers of caatinga. The genus is closely related to* Pilosocereus.

3

The HOT SPOTS

WHERE THE GREAT CACTI ARE

To know them [cacti], it is necessary to travel to the landscapes where they grow; it is impossible to appreciate them without visiting, among others, the spectacular area around Zapotitlán de las Salinas near Tehuacán, the blazing hot Cañon de Tomellín, the wild Sierra de la Mixteca, the profound gorge of Metztitlán in Hidalgo and of the Infiernillo in Querétaro, the basin of the Río Balsas, the high central plateau, the Pacific slopes of the western states, the Altar Desert, and the marvelous Baja California peninsula and the adjacent islands.

—HELIA BRAVO-HOLLIS,
LAS CACTÁCEAS DE MÉXICO

WHAT THE GREAT MEXICAN CACTOLOGIST Bravo-Hollis said about Mexican cacti is even truer of the giants of the cactus world. To appreciate them and to understand why their conservation is so urgently important, one must see them in their homes, must note firsthand how they evolved and fit into their landscapes, and must observe how native peoples adapt their way of life to them.

To see the greatest displays of great cacti, as Bravo-Hollis recommended, and see firsthand their practical uses, it is useful to be able to travel into Latin America and the Caribbean as well as to Arizona in the United States. In this chapter, I discuss what I consider to be the most impressive accessible locations where one can see them in their native lands. I rank these locations generally according to the

visual display, but also in terms of how humans have come to rely on the great cacti. All except for the Río Marañon region in Peru and Montevideo Canyon in Baja California are accessible with relative ease from commercial airports. Four of the sites lie within areas in close association with the Valle de Tehuacán, Puebla. Other sites not listed may be equally worthy, such as the arid cactus-rich valley enclaves of Columbia and Venezuela and the isolated Limeta Valley in southern Bolivia, but I have not visited the former and have been unable to identify some of the cacti I found in the latter, so I rank these spots lower on the list than those mentioned here. The hot spots described are ethnobotanically skewed and do not necessarily represent ranked areas of cactus evolution.

ARUBA CURAÇAO
BONAIRE

VENEZUELA

COLOMBIA

ECUADOR

Balsas
Cajamarca

Rio Marañón

PERU

Lima

Ayacucho

Cuzco

Colca
Canyon

Cotahuasi
Canyon

Arequipa

La Paz

Arica

Antofagasta

Parque Pan
de Azucar

Copiapó

Parque Llanos
de Carrizo

Santiago

CHILE

ARGENTINA

BRAZIL

CEARÁ

Recife

BAHIA

Caetité

Salvador

Itaobim

Rio de Jequitinhonha

MINAS
GERAIS

Rio de Janeiro

BOLIVIA

Sucre

Potosí

Tupiza

Salar de Uyuni

Humahuaca

Tilcara

JUJUY

SALTA

Quebrada
de Toro

PARAGUAY

URUGUAY

Buenos Aires

█ Columnar cacti
  hot spots

Figure 3.1. Chicos (Pachycereus weberi) *on hillside, San Bernardo, Río Acatlán valley, Puebla, Mexico.*

The Río Acatlán Valley, Puebla, Mexico

This well-populated and heavily used valley is located roughly 80–145 km south-southwest of the city of Puebla (which has an airport). The valley, some 65 km long, grades from southeast to northwest. It is home to at least eleven species of columnar cacti, all of which are used in one way or another. Although none of the cacti is endemic to the canyon, I know of nowhere else that cacti play such a prominent role in ethnobotany as in this canyon/valley. Great cacti are everywhere, dominating the natural landscape and often the human landscape as well. A well-used highway winds through the valley, making access easy to orchards of *Stenocereus* species (xoconochtlis, pitayos), and *Pachycereus weberi* (chicos). *Eschontrias* (jiotillas) are so heavily harvested that I suspect they are semidomesticated. In the southern portion of the valley, *Isolatocerei* (malayos) of immense size crowd the hillsides, intermingled

with *Polaskia chichipes*. Only slightly farther north, the chicos grow larger than anywhere else, and, although not as tall as plants in the Valle de Cuicatlán, they are colossal, perhaps the largest cacti in the world. *Neobuxbaumia mezcalaensis* (gigantes) routinely reach 15 m in height. *P. marginatus* (malinches), *P. pecten-aboriginum* (etchos), *Myrtillocacti* (garambullos), and the elusive *Pachycereus grandis,* rumored to be the tallest of all cacti, are there as well, all in stunning display. (Fig. 3.1.)

Reyes Metzontla and Environs, Valle de Zapotitlán, Valle de Tehuacán, Puebla, Mexico

Reyes Metzontla, a delightful though impoverished Popolocan town some 40 km southwest of Tehuacán, seems to have evolved with cacti. At least eleven species of columnars

Figure 3.2. Mitrocereus fulviceps *(left) and* Neobuxbaumia macrocephala *(right) near Reyes Metzontla, Puebla.*

are to be found within 10 km, including two extremely rare species, the chende *(Polaskia chende)* and the distinctive órgano de cabeza roja *(Neobuxbaumia macrocephala)*. The less rare but highly restricted chichipe *(Polaskia chichipe)* is also abundant. Columnar cacti grow in nearly every yard, several species in most. Inhabitants are intimately familiar with the various species, and most people can identify all. At least five species are cultivated for their fruits, and all eleven have some use or other. The outskirts feature vast tetecheras—groves of columnar cacti, mostly *Neobuxbaumia tetetzo*. Robust órgano de cabeza amarilla *(Mitrocereus fulviceps)* are intermingled with other giants in a most agreeable display. The noncolumnar cactus vegetation alone—especially the *Echinocactus platyacanthus* (silla de suegra, "mother-in-law's chair") and the *Beaucarnea gracilis (pata de elefante,* "elephant's foot")—makes the trip worthwhile. (Fig. 3.2.)

La Cañada de Cuicatlán and Environs, Oaxaca

This valley and canyon of the Río Grande Quiotepec, located roughly midway between Tehuacán, Puebla, and the city of Oaxaca, supports a profusion of columnar cactus species in a variety of habitats ranging from tropical deciduous forest to rangy thornscrub. It boasts fine stands of jiotillas, chicos, tetechos, babosos *(Pachycereus hollianus)*, and on a remote, lofty terrace overlooking a deep side canyon a singularly handsome grove of large chicomejos *(Pachycereus grandis)*. The arid, though well-watered, Valle de Cuicatlán is home to orchards of mango, mamey, and *chico zapotes.* Coffee trees flourish on high mesas above the valley. Excellent growth of *Mitrocereus fulviceps* extends well into the oak zone at nearly 1900 m on the southern reaches of the canyon. (Fig. 3.3.)

Figure 3.3. Pachycereus weberi, *near Quiotepec, Oaxaca. Mexican researchers believe this canyon contains the tallest cacti in Mexico.*

Figure 3.4. Balsas, Río Marañon, Peru. The village nestles on the bank of the great river. In the surrounding hills grow several varieties of columnar cacti.

The Río Marañon near Balsas, Amazonas, Peru

Balsas, a listless, isolated town in Peru (one day's drive east of Cajamarca by dirt road) is situated in an arid valley between Andean peaks that tower higher than 3000 m above the town. The Río Marañon is a major tributary of the Amazon, so finding a desert along its vast reaches is surprising. The Balsas area, including the steep slopes above, supports at least eight varieties of columnar cacti, including several endemics, especially *Browningias, Espostoas,* and the evolutionarily important *Calymmanthium.* Upstream and downstream from Balsas, the variety is enhanced with additional columnars in this cactus-rich desert region. If the region is extended 160 km downstream to include the region of Bagua, at least five additional species enhance the list. (Fig. 3.4.)

The Northeast Coast of the Gulf of California—Seri Country

Only five species of columnar cacti frequent this very dry desert coast area, none of them endemic. Still, cacti are numerous, especially the often gigantic cardón sahueso *(Pachycereus pringlei)*, the Seris' most important plant. The Seri people made (and to some extent still make) extensive use of all five species. This mountainous desert is blessed with copious dew, giving rise to a lushness and abundance of desert plants surprising for an area that receives less than 100 mm annual precipitation and may go for more than a year without rainfall. The sahuesos occur in large numbers on the bajadas. Of all Mexican indigenous peoples, Seris retain the most of their aboriginal language and traditional knowledge. Their familiarity with plants is astonishing to this day. (Fig. 3.5.)

Figure 3.5. Circle of sahuesos (Pachycereus pringlei), *El Cardonal, Sonora, possibly deliberately planted by Seris.*

Figure 3.6. Yatos (Stenocereus griseus) *growing on limestone, Aruba.*

The ABC Islands of the Caribbean—Aruba, Bonaire, and Curaçao

These three islands are home to only three species of columnar cacti, but the cacti occur in forests, especially on Bonaire, where a national park protects sensational groves of large kadushi *(Cereus repandus)* and *dadu (Pilosocereus lanuginosus)*. Elsewhere on the islands dense forests dominated by yatos *(Stenocereus griseus)* provide a spectacular experience. Many miles of cactus fences cordon off properties from one another. Older residents formerly made a vegetable stew based on the flesh of the kadushi. The cacti are abundant on all three islands. (Fig. 3.6.)

Montevideo Canyon, Baja California

This valley (despite its name, it is hardly a canyon) near Bahía de los Angeles on the Gulf of California supports stands of sahuesos *(Pachycereus pringlei)* and boojum trees *(Fouquieria columnaris)* of astonishing size intermingled with other strange plants, often nestled among granite boulders. Occasional individual sinitas *(Pachycereus [Lophocereus] schottii)*, pitayas agrias *(Stenocereus gummosus)*, and pitayas *(S. thurberi)* grow scattered among the giants. For sheer otherworldly beauty, this region of Baja California is unsurpassed. The canyon is uninhabited now, but was once regularly visited by the ancient inhabitants of Baja California. (Figs. 3.7 and 3.8.)

Salar de Uyuni and Environs, Bolivia

Though only two species are found around the world's largest and highest salt lake *(Trichocereus atacamensis, Oreocereus celsianus)*, the setting is of vast, unearthly beauty. The pasacanas flourish to nearly 4000 m, higher than 13,000 ft., elevation, enduring bitter cold with enthusiasm. An

Figure 3.7. Sahuesos (Pachycereus pringlei) *and* cirios *(boojum trees,* Fouquieria columnaris)*, near Bahía de los Angeles, Baja California.*

especially impressive growth can be found on the island called Isla Pescadores. In the lower (3000 m) Limeta Valley to the east are impressive numbers of the massive *T. werdermannianus* and the smaller but prolific *T. tacaquirensis*. Indigenous Quechua people view the cacti as part of their subsistence. (Fig. 3.9.)

The Balsas Depression, Cañon Infiernillo, Michoacán, Mexico

This hot basin extending to the Pacific Ocean supports at least eight species, most of which are used by natives. Although overgrazing is distressing, human population density is low for the most part, and the forests of cacti make for spectacular viewing, especially the abundant and giant pitires *(Stenocereus quevedonis)*. Decent highways make the area accessible, if Spartan in its tourist facilities. Most notable are forests of pitires and the unmistakable appearance of the attractive brown and gold cephalia of *Backebergia militaris*, the tiponchi. Around the village of Palo Pintado, an array of species of great size, including the pachón *(Stenocereus chrysocarpus)* and tepamo *(Pachycereus tepamo)*, makes for captivating viewing. High above the highway on limestone outcroppings grows a species of *Neobuxbaumia*, probably *tetetzo*. (Fig. 3.10.)

Figure 3.8. Pachycereus pringlei *with boojums, Montevideo Canyon, Baja California.*

Figure 3.9. Oreocereus celsianus *and* Trichocereus atacamensis, *near Salar de Uyuni, Bolivia, elevation around 3850 m (13,000 ft.).*

Figure 3.10. Pitireal (forest of Stenocereus quevedonis *trees) near Infiernillo, Michoacán. Bluish-tinted cacti in the center of the photo are* Pachycereus tepamo.

Figure 3.11. Trichocereus atacamensis, *Quebrada del Toro, Salta, Argentina. Elevation 2900 m (9700 ft.).*

Quebrada del Toro, Salta, Argentina

Only two species of columnars grow in the Quebrada del Toro: *Trichocereus atacamensis* (pasacanas) and *T. terscheckii* (cardones), but their habitats overlap, and where they meet, large numbers of them tend to hybridize. Above the conjunction, pasacanas grow in fine displays in high, cold desert featuring strange and varied topography. Below, the great cardones mix in the great trees of the yungas, semitropical cloud forests. Very old villages persist along the margins of the narrow valley. Crystal clear air and weird geological displays, including spectacular faulting, add to the wonder of the long canyon that ranges from below 1400 m to well above 3200 m elevation. (Fig. 3.11.)

Around Arequipa, Peru

Within a 100-km radius, mostly of very dry, high desert with piercingly clear air, at least six species of columnar cacti are to be found, species of *Armatocereus, Browningia, Neoraimondia, Corryocactus,* and *Weberbauerocereus.* An expanding metropolis relentlessly encroaches into the desert, but vast expanses remain where cacti are the dominant plants. Though none is massive, the variety is impressive, and their ability to colonize hyperarid landscapes is a tribute to cactus survival. In the accessible (with difficulty) but very remote Cañon Cotahuasi, fine stands of three additional species frequent the canyon sides. (Fig. 3.12.)

Figure 3.12. Weberbauerocereus weberbaueri *and* Browningia candelaris *on granite, near Arequipa, Peru, elevation near 2700 m (9000 ft.).*

The Environs of Alamos, Sonora, Mexico

Within a 40-km radius of the graceful colonial village Alamos, one is able not only to view seven species of columnar cacti, but also to amble through dense forests of pitayas. Etchos are especially useful vegetation indicators, exceeding the height of their tallest competing plants in thornscrub, while being surpassed by trees in tropical deciduous forest. On basaltic slopes of low mesas grow healthy populations of saguaros, and near the mining village of La Aduana are the robust sahuira *(Stenocereus montanus)* and occasionally the bearded pitaya barbona *(Pilosocereus alensis)*. Within the thornscrub at lower elevations are sinas *(Stenocereus alamosensis)* and sinitas. (Fig. 3.13.)

Figure 3.13. Etchos (Pachycereus pecten-aboriginum), *left side, and sahuira* (Stenocereus montanus) *near Potrero de Alcántara, Sonora.*

Figure 3.14. Cephalocereus totolapensis, *rainy season, near Totolapan, Oaxaca.*

Figure 3.15. Cephalocereus totolapensis, *dry season, near Totolapan, Oaxaca. Elevation around 1160 m (3500 ft.).*

The Environs of Totolapan, Oaxaca

Midway between Oaxaca City and the Isthmus of Tehuantepec, this canyon country is rich with at least eight species of columnars. Especially notable are huge (10 m tall) *Myrtillocactus schenckii* plants, the rare brujo *(Cephalocereus totolapensis)*, and the glittering viejita *(Pilosocereus quadricentralis)*. Impressive groves of very tall tetechos are found as well. Some of the densest forests of *Escontria* (here called *xuegos*) grow in the region. (Figs. 3.14 and 3.15.)

Teotitlán del Camino, Oaxaca

This town lies at an important junction of highways, some 40 km north of Cuicatlán, Oaxaca, another superb location for columnar cacti. Near Teotitlán are forests of tetechos so dense as to be impenetrable, in staggering numbers. *Cephalocereus columna-trajani* march on steep hillsides,

and great forests of babosos and chicos cover the lowlands. Xoconochtlis *(Stenocereus stellatus)* and pitayos de mayo *(Stenocereus pruinosus)* are also abundant, and the viejita *(Pilosocereus chrysacanthus)* is usually found in hillside assemblages. The town lies near the Oaxaca-Puebla state line and draws heavily from many native cultures in the region. (Fig. 3.16.)

Barranca de Metztitlán, Hidalgo, Mexico

This canyon-valley drains part of the high plateau and mountain region north of Mexico City, ranging from pine and oak forest at the top to very dry desert on the valley floor. Agriculture long ago supplanted the original vegetation of the flat bottomlands, but superb forests of malayos *(Isolatocereus dumortieri)* cover the higher canyon slopes. On steep hillsides nearer the bottom are numerous viejitos *(Cephalocereus senilis)* in a most agreeable assemblage of

Figure 3.16.
Tetechera (grove
of tetechos,
Neobuxbaumia
tetetzo), *Calipan,*
Puebla.

Figure 3.17. Isolatocereus dumortieri, *Barranca de Metztitlán, Hidalgo.*

desert plants. On a few limestone tallis slopes grow the rare totem pole cacti *(Neobuxbaumia polylopha)* in locations that are known only to locals and a few outsiders. The valley also supports *Stenocereus griseus, Myrtillocactus geometrizans,* and *Pachycereus marginatus,* the latter often planted for fences. Fresh pulque, the fermented juice of *Agave salmiana,* is sometimes sold along the roadside. (Fig. 3.17.)

NOTES

CHAPTER 1

1. See Yetman 1996, 2002; Yetman and Van Devender 2002.
2. Wallace 2002:44.
3. I base this figure, which I consider conservative, on my own experiences and on data published in Fleming 2002:208.
4. Lamb 1991:2. I am skeptical of the author's claim.
5. Salak 2000.
6. Hales and Rawley 1985.
7. Martin et al. 1998:282. Gentry's claim is surely credible. I have photographed a sahuira at least 15 m (50 ft.) tall, and local people assure me they grow taller.
8. Heywood 1993:6.
9. Vega-Villasante et al. 1995.
10. Fleming and Holland 1998.
11. *Pachycereus marginatus, Polaskia chende, P. chichipe, Stenocereus fricii, S. griseus, S. pruinosus, S. queretaroensis, S. stellatus,* and *S. treleasei.* The prickly pear most often cultivated for its fruits and tender pads is *Opuntia ficus-indica.* Elsewhere in Mexico, columnars propagated include *Pachycereus pecten-aboriginum, Stenocereus* sp. of coastal Oaxaca, *S. montanus, S. quevedonis,* and *S. thurberi.*
12. MacNeish 1967.
13. Emmart 1940:plate 28, also p. 231.
14. Terrazas Salgado and Mauseth 2002:29.
15. Nerd, Ravel, and Mezraki 1993; Mizrahi, Nerd, and Nobel 1997; Nerd, Tel-Zur, and Mizrahi 2002.
16. See, for example, Casas et al. 1997.
17. Mauseth, Kiesling, and Ostolaza 2002.
18. Ecologist Forrest Shreve (Shreve and Wiggins 1964) defined the eastern limit of the Sonora Desert as the isobar separating areas receiving more than 325 mm (13 in.) of annual rainfall from those receiving less. His map defining the area was much more liberal than his definition, including desert areas with annual rainfall of nearly 475 mm (20 in.).
19. Of the deserts of the United States, only the Sonoran has columnar cacti. The Mohave is apparently both too cold and too dry in the summer; the Great Basin and Chihuahuan deserts are too cold; and the Colorado (extreme southern California) is too dry.
20. Valiente-Banuet and Godínez-Álvarez 2002.
21. See Steenbergh and Lowe 1977.
22. See Felger, Johnson, and Wilson 2001:124.
23. Lau 1994.
24. Shreve 1935.
25. Ray Turner, personal communication, 1995.
26. Roberto Neumann, personal communication, 2001. Neumann believes that dating pasacanas is simple, for, he suggests, they develop one areole per rib each year. Thus, by counting the number of areoles on a rib, one can determine the age of the plant. This system works only for ribs that begin at ground level, however, because the areoles on branches will date from the initiation of the branch.
27. Cornejo 1994:129.
28. Cornejo 1994. Valiente-Banuet and associates (2002) attribute seventy species of columnar cacti to Mexico, but I am unable to verify more than sixty, and that number may be excessive apart from species undescribed as of 2002—of which at least three remain. The discrepancy is another symptom of the unending battle between lumpers and splitters and of disagreement as to what constitutes a columnar cactus.
29. Braun and Esteves Pereira 2001.
30. The tribes are Cacteae and Pachycereeae in Greater Mexico; Browningieae, Notocacteae, and Trichocereeae in the Andes; and Cereae in Brazil (Anderson 2001:40). A tribe is a taxonomic subcategory of the family. The order of scope runs: family, tribe, genus, species.
31. Madsen 1989.
32. Mourelle and Ezcurra 1997.
33. *Cephalocereus columna-trajani, Neobuxbaumia macrocephala, Pachycereus hollianus, Polaskia chende,* and *P. chichipe.*
34. *Neobuxbaumia squamulosa, Stenocereus chrysocarpus,* and *S. fricii.*
35. *Armatocereus brevispinus, A. godingianus,* and *Espostoa frutescens.*
36. *Armatocereus arduus, A. laetus, A. rauhii, Browningia altissima, B. chlorocarpa, B. microsperma, B. pilleifera, Espostoa blossfeldiorum, E. laticornua, E. mirabilis, E. superba, E. utcubambensis.* In fact, all or nearly all the columnar cacti of the Río Marañon region appear to be more or less narrowly endemic. Further studies there will surely turn up additional species.
37. Pérez de Ribas 1999:88.
38. Pfefferkorn 1989:75–76.
39. Turner, Bowers, and Burgess (1995) observe that *Pachycereus pringlei,* the sahueso (or cardón, as it is called in Baja California), flourishes on the Baja California peninsula, where most of the moisture is derived from winter storms. They hypothesize that recruitment probably depends on monsoonal moisture, which is notoriously unreliable except in the southern quarter of the peninsula. In the heart of cardón habitat on Baja California, entire summers may pass with no rainfall.

40. Gibson and Nobel 1986:74. The primitive cacti of the genus *Pereskia* combine elements of C_3 (the pathway of carbon in photosynthesis utilized by most nontropical plants) and CAM, but appear to be closer to C_3 (Gibson and Nobel 1986:78–86). Many tropical grasses or temperate grasses of pantropical origin use what is known as C_4 metabolism or pathway of photosynthesis. Outstanding examples of C_4s are corn and Bermuda grass, and typical C_3s are wheat and apples. Many succulents—including agaves, bromeliads, ballmosses, and sedums—use CAM. Nobel cites studies demonstrating that the amount of water lost to plants through water vapor passing from the plants to the air during hot days is dramatically higher than the loss under cooler temperatures at night. Water loss is 4.8 times higher at 20°C than at 5°C. "If the temperatures were 15°C cooler at night than during the daytime, which is realistic for deserts where columnar cacti occur, transpiration would be three- to fivefold lower at night than during the daytime . . . which underscores the importance of nocturnal stomatal opening and CAM for water conservation by columnar cacti" (2002, 194–195). Nobel and Bobich (2002) note that plants with C_3 metabolism include the majority of angiosperms—that is, flowering plants—of temperate climates. C_4 plants are angiosperms of tropical origin, including Bermuda grass, corn, sugarcane, and many noxious weeds of the tropics. The subscripts 3 and 4 refer to the number of carbon atoms in the molecules at critical stages of starch formation.

41. Bravo-Hollis 1978a.

42. Cornejo and Simpson 1997:1499; Nobel 2002:191.

43. See Gibson and Nobel 1986:14.

44. See, for example, Benson 1957; Barthlott and Hunt 1993.

45. Mabberly 1997:112.

46. Anderson 2001.

47. Cited in Mitich 1997:33.

48. Cornejo (1994) has argued that phylogenetically close members of the genus *Stenocereus* often have double (or half) the number of ribs of their closest relative.

49. Dubrosky and North 2002.

50. Shreve 1935.

51. Yet their greatest diversity and rate of endemism is in Mexico, with more than nine hundred species, more than six hundred of which are endemic (Challenger 1998).

52. Lowe and Steenbergh 1981.

53. Mauseth 1989.

54. Wallace and Gibson 2002.

55. In spite of the convergence, the similar climates also reveal startling disjunctions in desert plant species. *Larrea divaricata,* the common creosote of the southwestern United States, disappears in central Sonora, Mexico, not to reappear again until the arid valleys of northwest Argentina with four species of the genus and several of the closely related genus *Bulnesia.* The same is true of the Sonoran palo verde *(Parkinsonia praecox)* and the caper *(Atamisquea emarginata),* both common in coastal thornscrub and desertscrub in Sonora, but absent in the thornscrub from Sinaloa south. How these three species jumped more than 6000 or so km (3700 mi.) from south to north or north to south is an intriguing mystery.

56. A. Gentry 1982:557; for discussions of these theories, see also Leuenberger 1986; Mauseth 1989; and Van Devender 1990.

57. More specifically, the Caribbean-Columbian Cretaceous Igneous Province. See Kerr et al. 1996:111.

58. Weiner 1994.

59. Leuenberger 1986.

60. Mexico did not assume its present form until the Miocene (ca. 23–25 million years ago). Until that time, the Mexican terrane included only the eastern and southeastern parts of the current landmass (Steward 1992).

61. See discussions in Van Devender 1990, 2002; Soriano and Ruiz 2002:244–247.

62. Gibson et al. 1986; Wallace and Gibson 2002:10.

63. Gibson 1991a.

64. Wallace 2002.

65. Gibson et al. 1986; Wallace and Gibson 2002. Cornejo (1994) carried out similar studies, analyzing cactus morphology or anatomy.

66. Del Castillo 1996.

67. Gibson and Nobel 1986.

68. Valiente-Banuet and Godínez-Álvarez 2002:95.

69. Jorge Meyrán (1970) provided the first readily available account of the cacti of that valley complex. The associated valleys (including the valley of the Río Grande Quiotepec in Oaxaca) actually contain as many as nineteen species.

70. Yetman and Búrquez 1996.

71. Lamb 1991:52.

72. Bravo-Hollis 1978a.

73. See Buxbaum 1958.

74. Gibson and Horak 1978.

75. Wallace 2002.

76. Gibson and Horak 1978; Cornejo 1994.

77. Gibson and Nobel 1986:216.

78. For the Seri names of these species, I have used the method of transliteration adopted by Felger and Moser (1985) rather than my own notations, which were similar.

79. Aronson 1990.

80. Braun and Esteves Pereira 2005:87; see also Reyes Santiago et al. 2004.

CHAPTER 2

1. See Fontana 1980.

2. Baxter 1932.

3. Pauley 1984.

4. Steenbergh and Lowe 1977.

5. Nobel and Bobich 2002:67.

6. Steenbergh and Lowe 1976.

7. Hastings and Turner 1965.

8. Extrapolated from data for the region compiled by Hastings and Humphrey (1969).

9. See Steenbergh 1972:419.

10. Crosswhite 1980.

11. Barthlott and Hunt 1993:190.

12. See especially Underhill 1946, Crosswhite 1980, and Hodgson 2001. See Fontana 1980 for a comprehensive bibliography.

13. Crosswhite 1980:5.

14. Selso Villegas, personal communication, 2003.

15. Hodgson 2001.

16. Bernard Fontana, e-mail, March 20, 2002.

17. Coll y Toste 1972. Dan Austin was kind enough to provide me with this reference.

18. Enriquena Bustamante, personal communication, February 2005. She measured 185 individual plants near Sonoita, Sonora, across the border from Organ Pipe Cactus National Monument, and 93 plants within the Pitayal.

19. Gibson 1989b.

20. Sahley 2001.

21. Barco 1980:149; see also Hodgson 2001:142.

22. Crosby 1994:213.

23. Arbelaez 1991.

24. Pérez de Ribas 1999:89.
25. Felger and Moser 1985.
26. Mercado and Granados 1999; Casas, Valiente-Banuet, and Caballero 2002.
27. Gibson 1989b.
28. Felger, Johnson, and Wilson 2001.
29. Pimienta-Barrios and Nobel 1994:79.
30. *Backebergia militaris, Escontria chiotilla, Neobuxbaumia mezca-laensis, N. squamulosa, Pachycereus tepamo, P. weberi, Pilosocereus alensis, (P. purpusii?), Stenocereus chrysocarpus, S. friçii, S. quevedonis,* and *S. standleyi.* There are almost certain to be *Pachycereus pecten-aboriginum* and perhaps *P. grandis* as well, although I have not seen any specimens. Upstream in Guerrero are the narrowly endemic *Neobuxbaumia multiareolata* and *Stenocereus zopilotensis.*
31. Sánchez-Mejorada 1972.
32. Pimienta-Barrios 1999.
33. Gibson 1989b.
34. Cornejo 1994.
35. Casas et al. 1997:288. The study does not specify whether the total included only fruits of *S. stellatus* or those of *S. pruinosus* as well. The two species are usually planted together in orchards and in fencerows. *S. pruinosus* produces fruits in spring and sometimes in fall, whereas *S. stellatus* produces fruits in August and September. See also Casas et al. 1999.
36. Bravo-Hollis 1978a:584; Anderson 2001:646.
37. Casas, Valiente-Banuet, and Caballero 2002. Gibson (1991b), in his monograph on the genus, has a photo of a domestic *S. griseus* from an unidentified location in southern Puebla. If the location is the valley of the lower Río Acatlán (around 1000 m [3400 ft.]), it could well be the typical habitat for *S. griseus*. I have not managed to locate *S. griseus* in Puebla. It is to be expected in the low valleys in the eastern portion of the state. See also Gibson 1991c.
38. Casas, Caballero, and Valiente-Banuet 1999; see also Casas et al. 1997:285, 290.
39. Cornejo 1994.
40. Cornejo 1994.
41. Gibson 1988d.
42. Bravo-Hollis 1978a.
43. Anderson 2001.
44. Arreola Nava and Terrazas 2004. As is often the case when new species are described, the new nominees tend to be lacking in ethnobotanical information.
45. Sánchez-Mejorada 1970:33.
46. Yetman and Búrquez 1996.
47. Felger and Moser 1985.
48. Gibson et al. 1986; Gibson 1989a.
49. Bashan, González-Bashan, and León de la Luz 2001.
50. Barco 1980:158.
51. Salak 2000.
52. Turner, Bowers, and Burgess 1995.
53. For discussion of polysexuality of sahuesos, see Fleming, Nuñez, and Sternberg 1993; Murowski et al. 1994.
54. Felger and Moser 1985:254.
55. Shreve 1935.
56. Yetman and Búrquez 1996.
57. Nabhan 2000.
58. Bowen 2000.
59. Hodgson 2001:132.
60. Felger and Moser 1985.
61. Angelina Martínez Yrízar, personal communication, 2000.
62. Fleming 2002.
63. Griffen 1959.

64. Moran 1997.
65. Steenbergh and Lowe 1977:188–207; Bashan, González-Bashan, and León de la Luz 2001.
66. Ray Turner, personal communication, 1997.
67. Turner, Bowers, and Burgess 1995:307; see also Hastings and Turner 1965.
68. Bashan et al. 2000.
69. Sobarzo 1961.
70. Cornejo notes that *Pachycereus pecten-aboriginum* is the only columnar cactus that crosses what he labels regions 1–4 of Mexico (1994:29); see also H. Gentry 1942:53.
71. H. Gentry 1982.
72. Bravo-Hollis 1978a:68.
73. Herrera and Calderón Villagómez 1994; Casas et al. 1999.
74. Arias, Terrazas, and Cameron 2003.
75. Gibson and Nobel 1986:218.
76. Bravo-Hollis 1978a.
77. Beutelspacher and Ramírez Martínez 1973.
78. Gibson and Nobel 1986:216.
79. Gibson and Horak 1978; Anderson 2001.
80. Fleming and Holland 1998.
81. León de la Luz and Fogel 2005.
82. See Lavender 1989.
83. Felger and Moser 1985:248.
84. Gama-López and Arias 1998.
85. Wallace 2002.
86. Martínez-Alvarado and Flores-Castorena 1997.
87. Elevation and rainfall figures are from Cornejo 1994:384–385.
88. Thomas Van Devender, personal communication, 2005.
89. Valiente-Banuet and Ezcurra 1991; Valiente-Banuet et al. 1996.
90. Casas, Caballero, and Valiente-Banuet 1999; Valiente-Banuet, Vite, and Zavala-Hurtado 1991.
91. Scheinvar and Sánchez-Mejorada 1990.
92. Bravo-Hollis 1978a.
93. Bravo-Hollis 1978a.
94. Bravo-Hollis, Scheinver, and Sánchez-Mejorada 1973.
95. Lau 1994.
96. Lau 1994.
97. Bravo-Hollis 1978a.
98. Bravo-Hollis 1978a.
99. Lau 1986.
100. Greenwood 1964.
101. Haselton 1942.
102. Braun and Esteves Pereira 2002.
103. Zappi 1994; Braun and Esteves Pereira 2002.
104. Anderson 2001.
105. Cornejo and Simpson 1997.
106. Flores Martínez et al. 1991.
107. Gibson 1988a, 1988b, 1988c.
108. Casas and Barbera 2002:148. The category of edible seeds requires that the seeds be separated from the pulp and eaten separately, as opposed to simply being eaten when the fruit is consumed.
109. Carmona 2001.
110. This terminology is used by Casas and associates (1997) for wild plants assimilated into domestic production.
111. Gibson 1991c.
112. Hunt, Taylor, and Charles 2006.
113. Anderson 2001. See also Hunt 1991, 1992; Hunt, Taylor, and Charles 2006.
114. Graham Charles, personal communication, 2006.

115. See Sánchez-Mejorada 1973 for the history of the plant's taxonomy.

116. See Mauseth et al. 2005.

117. Axel Nielson, personal communication, 2001.

118. Roberto Neumann, personal communication, 2001.

119. Roberto Neumann, personal communication, 2001.

120. Fernández Distel 1997.

121. However, Rocha noted the special water-catchment system designed to supplement the natural rainfall arriving at the plants' roots.

122. Méndez 2000.

123. For example, Arenas and Scarpa (1998) do not list the name *unquillo*. Most of their study was carried out among indigenous peoples in Argentina and Paraguay. They report several different names in indigenous languages.

124. Wallace 2002.

125. Charles and Woodgyer 2003.

126. See Charles 2004.

127. Madsen 1989.

128. See Charles 1999.

129. Charles 1999.

130. Piacenza and Ostolaza 2002.

131. Ostolaza, Mitich, and King 1985.

132. Bustamante and Búrquez 2005.

133. Mauseth, Kiesling, and Ostolaza 2002:156–159.

134. Cáceres et al. 2000.

135. Wallace 2002.

136. Madsen 1989.

137. Ostolaza 1984, 2001.

138. Sharon 1972.

139. Rudgley 1998.

140. Arbelaez 1991.

141. Ostolaza 2001:35.

142. Burger 1992.

143. Ostolaza 2001.

144. Joralemon and Sharon 1993.

145. Wallace and Gibson 2002.

146. Quimi and Stothert 1994.

147. Madsen 1989.

148. Jens Madsen, personal communication, 2002.

149. Anderson 2001.

150. Anderson 2001.

151. Aronson 1990; Gibson 1991a. The entire coastline of the Atacama—from central Chile to northern Peru—receives moisture in the form of *comanchaca* (Chile) or *garúa* (Peru): a cold, drizzly fog. The reliability and density of this moisture source, not easily measurable in rain gauges, gives rise to specific coastal plant communities. It also makes for living conditions considered rather depressing by coastal human inhabitants, who must deal with months of nearly unending fog and chilly temperatures. Consider the travails of women who must wash and dry the family laundry where there is precious little running water and a saturated gloomy atmosphere that barely permits laundered clothes to dry. During the garúas, Lima, Peru, although hardly 10° south of the equator, has a climate reminiscent of that of London in the winter.

152. Ostolaza 1986.

153. Charles 2000.

154. Ostolaza 1997.

155. Yetman 2004.

156. Mauseth, Kiesling, and Ostolaza 2002.

157. Nerd, Ravel, and Mezraki 1993.

158. Morton 1967.

159. *P. arribidae* and *P. brasiliensis* range as far south as the Tropic of Capricorn. *P. alensis* grows far north of the Tropic of Cancer.

160. Zappi 1994.

161. Zappi 1994.

Glossary and Common Names

Locations in parentheses indicate where a given common name is used.

aaqui: Mayo term for *Stenocereus thurberi* (Sinaloa, Sonora)

acompes: *Pachycereus hollianus*

ahuacullo: *Echinopsis pachanoi,* also called the San Pedro cactus

apple cactus: *Cereus hildmannianus* (United States)

areole: wartlike growth on cactus ribs from which spines and flowers grow

baboso: *Pachycereus hollianus;* means "drooler" in Spanish

bacate: stick or pole for gathering cactus fruits

bajada: gentle slope below mountains, consisting of eroded material

brujo: *Cephalocereus totolapensis;* means "witch" in Spanish

cabeza de viejo: *Pilosocereus purpusii* and *P. leucocephalus*

Cáhita: the language spoken by Mayos and Yaquis

calcareous: rich in calcium carbonate; caliche or limestone derived

Californio: Baja Californian of many generations

CAM: Crassulacean acid metabolism; process used by cacti and other succulents to convert carbon dioxide to food

cardón: a common Spanish name given a variety of columnar cacti; means "big thistle"

caripari: *Neoraimondia herzogiana*

cereoid: (noun or adjective) columnar, columnar cactus

chende: *Polaskia chende*

chichipe: *Polaskia chichipe*

chico: *Pachycereus weberi* (southwestern Puebla); means "little" in Spanish

chicomejo: *Pachycereus grandis* (Puebla)

cladode: a spiny segment on a cactus that is detachable and thus aids in propagation

clavija: *Neobuxbaumia mezcalaensis*

cochal: *Myrtillocactus cochal* (Baja California)

compés: *Pachycereus hollianus* (western Puebla)

conoxtli: *Escontria chiotilla* (southwest Puebla)

copao: *Eulychnia acida* (Chile)

creeping devil: *Stenocereus eruca* (Baja California)

cuepetla (Mixtec): *Stenocereus pruinosus*

dadu: *Pilosocereus lanuginosus* (Netherlands Antilles)

etcho: *Pachycereus pecten-aboriginum* (Sonora)

faica: *Browningia microsperma*

garambullo: *Myrtillocactus geometrizans* (usually); also *M. schenkii*

gigante: giant; *Neobuxbaumia mezcalaensis* (Puebla)

glochids: tiny spines found on *Opuntias* and other cacti

Gondwana: the massive continental mass of the Paleozoic period

grenadier's cap: *Backebergia militaris* [= *Pachycereus militaris*]

huarango: *Weberbauerocereus* sp. (Arequipa, Peru)

inferior ovary: female structures anchored below male structures in flowers

jiotilla: *Escontria chiotilla* (Puebla)

kadushi: *Cereus repandus* (Netherlands Antilles)

kehuaylla: *Trichocereus atacamensis*

malinche: *Pachycereus marginatus;* name of Cortes's interpreter-mistress

meristem (apical meristem): the growth tip of a plant

malayo: *Isolatocereus dumortieri* (Oaxaca, Puebla)

mandacarú: *Cereus jamacaru*

monte: the bush, wild country

morphology: pertaining to structure

nochtli: *Neobuxbaumia squamulosa* ("cactus" in Hahuatl)

órgano: name for a wide variety of columnars; means "organ"

órgano de cabeza amarilla: *Mitrocereus fulviceps* (Puebla); means "yellow-headed organ"

órgano de cabeza roja: *Neobuxbaumia macrocephala* (Puebla); means "red-headed organ"

pachón: *Stenocereus chrysocarpus* (Michoacán)

padre nuestro: *Myrtillocactus geometrizans*

pasacana: *Trichocereus atacamensis,* also the fruit thereof (Argentina, Chile)

photosynthesis: the process whereby plants use sunlight as the energy source to produce new growth

phylogeny: evolutionary development

pitaya, pitayo: *Stenocereus thurberi;* several *Stenocereus* species (Mexico) and their fruits; sometimes generic for columnar cactus fruit

pitaya agria: *Stenocereus gummosus* (Baja California, Sonora); means "tart pitaya"

pitaya barbona: *Pilosocereus alensis* (Sinaloa, Sonora); means "bearded pitaya"

pitaya dulce: *Stenocereus thurberi* (Sonora); means "sweet pitaya"

pitaya marismeña: *Stenocereus standleyi* (Sinaloa, Guerrero), means "maritime pitaya"

pitayal: pitayo forest; *Stenocereus chacalapensis* (Oaxaca)

pitayo de aguas: *Stenocereus friçii* (Michoacán); means "pitayo of the rains"

pitayo de mayo: *Stenocereus pruinosus; Neobuxbaumia mezcalaensis* (Oaxaca, Puebla)

pitayo de octubre: *Stenocereus stellatus* (Oaxaca, Puebla)

pitayo de Querétaro: *Stenocereus queretaroensis*

pitire: *Stenocereus quevedonis* (Michoacán)

Pleistocene epoch: roughly the past 2 million years, ending with the last glacial age

quiotilla: *Escontria chiotilla*

rabo de zorro: *Espostoa calva, E. mirabilis*

saguaro: *Carnegiea gigantea* (Arizona, Sonora); *Azureocereus hertlingianus* (Peru)

sahueso: *Pachycereus pringlei* (Sonora)

sahuira: *Stenocereus montanus* (Sinaloa, Sonora)

salea: fruit of the *Neobuxbaumia tetetzo* (Puebla)

sancayo: *Corryocactus brevistylus*

sanquey: *Trichocereus cuzcoensis* (Peru)

San Pedro cactus: *Echinopsis pachanoi* (Ecuador, Peru)

sapang haurnis: *Neoraimondia arequipensis;* means "lonely woman" in Quechua

sinita (senita): *Pachycereus [Lophocereus] schottii* (Arizona, Mexico)

sina: *Stenocereus alamosensis* (Sonora)

sinaaqui: Mayo term for rare columnar, perhaps a hybrid of *Stenocereus alamosensis* and *S. thurberi*

stamens: male structures in flowers, usually very numerous in cacti

soberbio: *Browningia candelaris* (Chile)

spongy parenchyma: tissue in stems that takes in water and releases it when the plant needs it

sympatric: sharing a habitat

tenchanochtli: *Pachycereus weberi* (Valle de Tehuacán)

tepamo: *Pachycereus tepamo* (Michoacán)

tetecho, tetetzo: *Neobuxbaumia tetetzo* (Oaxaca, Puebla)

tetechera: grove or forest of columnar cacti, especially tetechos, in southern Mexico

tiponchi: *Backebergia militaris* (Michoacán)

toothpick cactus: *Stetsonia coryne*

totem pole cactus: *Neobuxbaumia polylopha; Pachycereus [Lophocereus] schottii,* var. *monstruosus*

tunillo: *Stenocereus treleasei* (Oaxaca)

tunshichi (Mixtec): *Neobuxbaumia sanchezmejoradae*

ucle: *Cereus validus*

unquillo: *Stetsonia coryne* (Argentina)

vela: *Neobuxbaumia multiareolata*

vichisova: *Myrtillocactus schenkii*

viejita: *Pilosocereus chrysacanthus*

viejito: *Cephalocereus columna-trajani* (Mexico); *C. senilis; Pilosocereus* spp. (Mexico)

xoconochtli: *Stenocereus stellatus* (Oaxaca, Puebla)

xonochtli: *Escontria chiotilla* (Oaxaca, Puebla)

xuego: *Escontria chiotilla* (eastern Oaxaca)

yato: *Stenocereus griseus* (Netherlands Antilles)

yonco: *Browningia pilleifera*

zonca: *Espostoa calva, E. lanata, E. mirabilis*

REFERENCES

Anderson, Edward. 2001. *The cactus family.* Portland, Ore.: Timber Press.

Arbelaez, M. S. 1991. The Sonoran missions and Indian raids of the eighteenth century. *Journal of the Southwest* 33: 366–377.

Arenas, Pastor, and G. F. Scarpa. 1998. Ethnobotany of *Stetsonia coryne* (Cactaceae), the "cardón" of the Gran Chaco. *Haseltonia* 6: 42–51.

Arias, Salvador, Teresa Terrazas, and Kenneth Cameron. 2003. Phylogenetic analysis of *Pachycereus* (Cactaceae, Pachycereeae) based on chloroplast and nuclear DNA sequences. *Systematic Botany* 28: 547–557.

Aronson, James. 1990. Desert plants of use and charm from northern Chile. *Desert Plants* 10: 65–86.

Arreola Nava, H., and Teresa Terrazas. 2004. *Stenocereus zopilotensis:* A new species from Mexico. *Brittonia* 56: 96–100.

Barco, Miguel del. 1980. *Natural history of Baja California.* Translated by F. Tiscareno. Los Angeles: Dawson's Book Shop.

Barthlott, W., and David R. Hunt. 1993. Cactaceae. In K. Kulitzki, editor, *The families and genera of vascular plants,* 2:151–196. Hamburg: Springer Verlag.

Bashan, Yoav, Luis E. González-Bashan, and José Luís León de la Luz. 2001. King cactus: The giant cardón cactus of Baja California. *Wildflower* (winter): 11–16.

Bashan, Yoav, G. Toledo, Luis E. González-Bashan, and G. Holguín. 2000. La caida de los gigantes: Un analisis del decaimiento del cardón *(Pachycereus pringlei)* in Baja California Sur. *Ciencia y Desarrollo* 26: 30–37.

Baxter, E. M. 1932. California cacti; *Carnegiea gigantea*—giant cactus. *Cactus and Succulent Society of America* 3: 134–135.

Benson, Lyman. 1957. *Plant classification.* Boston: D. C. Heath.

Beutelspacher B. C., and M. Ramírez Martínez. 1973. Polinazación en *Stenocereus marginatus* (D.C.) Britton & Rose. *Cactáceas y Suculentas Mexicanas* 18: 80–84.

Bowen, Thomas. 2000. *Unknown Island: Seri Indians, Europeans, and San Esteban Island in the Gulf of California.* Albuquerque: University of New Mexico Press.

Braun, Pierre, and E. Esteves Pereira. 2001. *Kakteen und andere Sukkulenten in Brasilien.* Oldenberg, Germany: Isensee.

_____. 2002. *Pilosocereus goianis. British Cactus and Succulent Journal* 20: 93–102.

_____. 2005. The Melocacti of Chapada Grande, Brazil, and the conservation status of *Melocactus deinacanthus. Cactus and Succulent Journal (U.S.)* 77(2): 82–89.

Bravo-Hollis, Helia. 1978a. *Las cactáceas de México.* Vol 1. Mexico City: Instituto de Biología, Universidad Autónoma de México.

_____. 1978b. Consideraciones acerca de la clasificación, morfología y distribución de las cactáceas. *Cactáceas y Suculentas Mexicanas* 23: 9–21.

Bravo-Hollis, Helia, L. Scheinvar, and H. Sánchez-Mejorada. 1973. Estudio comparativo del género *Neobuxbaumia.* IV. *Neobuxbaumia multiareolata. Cactáceas y Suculentas Mexicanas* 18: 59–67.

Britton, Nathaniel L., and Joseph Rose. 1919–23. *The Cactaceae.* 4 vols. Publications of the Carnegie Institute of Washington no. 248. Washington, D.C.: Carnegie Institute.

Burger, Richard. 1992. *Chavín and the origins of Andean civilization.* New York: Thames and Hudson.

Bustamante, Enriquena, and Alberto Búrquez. 2005. Fenología y biología reproductiva de las cactáceas columnares. *Cactáceas y Suculentas Mexicanas* 50: 64–88.

Buxbaum, Franz. 1958. The phylogenetic division of the subfamily Cereoideae, Cactaceae. *Madroño* 14: 177–206.

Cáceres, Fátima, Antonio García, Elio Ponce, and Rafael Andrade. 2000. "El Sancayo," *Corryocactus brevistylus* (Schumann ex Vaupel) Britton y Rose. *Quepo* 14: 37–42.

Carmona, A. 2001. Variación morfológica en poblaciones silvestres, manejadas y cultivadas de *Polaskia chichipe* en el Valle de Tehuacán-Cuicatlán. Master's thesis, Universidad de Colima, Colima, Mexico.

Casas, Alejandro, and G. Barbera. 2002. Mesoamerican domestication and diffusion. In Park Nobel, editor, *Cacti: Biology and uses,* 143–162. Berkeley: University of California Press.

Casas, Alejandro, Javier Caballero, and Alfonso Valiente-Banuet. 1999. Use, management, and domestication of columnar cacti in south-central Mexico: A historical perspective. *Journal of Ethnobiology* 19: 71–95.

Casas, Alejandro, B. Pickersgill, Javier Caballero, and A. Valiente-Banuet. 1997. Ethnobotany and domestication in *xoconochtli, Stenocereus stellatus* (Cactaceae), in the Tehuacán Valley and La Mixteca Baja, México. *Economic Botany* 51: 279–292.

Casas, Alejandro, Alfonso Valiente-Banuet, and Javier Caballero. 2002. Evolutionary trends in columnar cacti under domestication in south-central Mexico. In Ted Fleming and Alonso Valiente-Banuet, editors, *Columnar cacti and their mutualists: Evolution, ecology, and conservation,* 147–163. Tucson: University of Arizona Press.

Casas, Alejandro, Alfonso Valiente-Banuet, Alberto Rojas-Martínez, and Patricia Dávila. 1999. Reproductive biology and the process of domestication of the columnar cactus *Stenocereus stellatus* in central Mexico. *American Journal of Botany* 86: 534–542.

Challenger, Anthony. 1998. *Utilización y conservación de los ecosistemas terrestres de México.* Mexico City: Universidad Autónoma de México.

Charles, Graham. 1999. The genus *Espostoa* Br. & R. *British Cactus and Succulent Journal* 17: 69–79.

_____. 2000. *Browningia candelaris* (Meyen) Br. & R. *British Cactus and Succulent Journal* 18: 39–42.

_____. 2004. Balsas, Peru—a dream come true. *British Cactus and Succulent Journal* 22: 2–9.

Charles, Graham, and Elizabeth Woodgyer. 2003. A new species of *Espostoa* from Peru. *British Cactus and Succulent Journal* 21: 69–74.

Coll y Toste, Cayetano. 1972. *Clásicos de Puerto Rico.* 2d ed. San Juan, Puerto Rico: Ediciones Latinoamericanas.

Cornejo, Dennis O. 1994. Morphological evolution and biogeography of Mexican columnar cacti, tribe Pachycereeae, Cactaceae. Ph.D. diss., University of Texas.

Cornejo, Dennis O., and B. Simpson. 1997. Analysis of form and function in North American columnar cacti (Tribe Pachycereeae). *American Journal of Botany* 84: 1482–1501.

Crosby, Harry. 1994. *Antigua California.* Southwest Center series. Albuquerque: University of New Mexico Press.

Crosswhite, Frank. 1980. The annual saguaro harvest and crop cycle of the Papago, with reference to ecology and symbolism. *Desert Plants* 2: 3–61.

Del Castillo, Roderigo. 1996. Ensayo sobre el fenómeno calcícola-calcuga en cactáceas mexicanas. *Cactáceas y Suculentas Mexicanas* 41: 3–11.

Dubrosky, Joseph, and Gretchen North. 2002. Root structure and function. In Park Nobel, editor, *Cacti: Biology and uses,* 41–56. Berkeley: University of California Press.

Emmart, Emily, translator and editor. 1940. *The Badianus Manuscript: An Aztec herbal of 1552.* Baltimore: Johns Hopkins University Press.

Felger, Richard, Matt Johnson, and Michael Wilson. 2001. *Trees of Sonora, Mexico.* Oxford: Oxford University Press.

Felger, Richard, and M. B. Moser. 1985. *People of the desert and sea: Ethnobotany of the Seri Indians.* Tucson: University of Arizona Press.

Fernández Distel, A. 1997. La "yista" del cardón pasacana (*Trichocereus pasacana* [Web.] Britton et Rose, Cactaceae) en la provincia de Jujuy, Argentina. *Parodiana* 10(1–2): 1–9.

Fleming, Ted. 2002. Pollination biology of four species of Sonoran Desert columnar cacti. In Ted Fleming and Alonso Valiente-Banuet, editors, *Columnar cacti and their mutualists: Evolution, ecology, and conservation,* 207–224. Tucson: University of Arizona Press.

Fleming, Ted, and J. N. Holland. 1998. The evolution of obligate pollination mutualism: Senita cactus and senita moth. *Oecologia* 114: 368–375.

Fleming, Ted, R. Núñez, and L. Sternberg. 1993. Seasonal changes in the diets of migrant and non-migrant nectarivorous bats as revealed by carbon stable isotope analysis. *Oecologia* 94: 72–75.

Flores Martínez, A., G. Manzanero Medina, R. Aguilar Santelises, and A. Saynes Vásquez. 1991. Importancia ecológica y económica de *Escontria chiotilla* (Weber) Rose en la porción este de las valles centrales de Oaxaca. *Cactáceas y Suculentas Mexicanas* 36: 16–22.

Fontana, Bernard. 1980. Ethnobotany of the saguaro, an annotated bibliography. *Desert Plants* 2: 63–78.

Gama-López, Susana, and Salvador Arias. 1998. Una nueva especie de *Pachycereus* (Cactaceae) del Occidente de México. *Novon* 8: 359–363.

Gentry, Alwin H. 1982. Neotropical floristic diversity: Phytogeographical connections between Central and South America. Pleistocene climatic fluctuation or an accident of the Andean Orogeny? *Annals of the Missouri Botanical Gardens* 69: 557–593.

Gentry, Howard S. 1942. *Río Mayo plants.* Washington, D.C.: Carnegie Institution.

_____. 1982. Sinaloan deciduous forest. In D. Brown, editor, *Biotic communities of the American Southwest—United States and Mexico.* Special issue of *Desert Plants* 4: 73–77.

Gibson, Arthur. 1988a. The systematics and evolution of subtribe Stenocereinae. 2. *Polaskia. Cactus and Succulent Journal (U.S.)* 60: 55–62.

_____. 1988b. The systematics and evolution of subtribe Steno-cereinae. 3. *Myrtillocactus. Cactus and Succulent Journal (U.S.)* 60: 109–116.

_____. 1988c. The systematics and evolution of subtribe Steno-cereinae. 4. *Escontria. Cactus and Succulent Journal (U.S.)* 60: 161–167.

_____. 1988d. The systematics and evolution of the subtribe Stenocereinae. 6. *Stenocereus stellatus* and *Stenocereus tre-leasei. Cactus and Succulent Journal (U.S.)* 61: 15–32.

_____. 1989a. The systematics and evolution of subtribe Steno-cereinae. 7. The machaerocerei of *Stenocereus. Cactus and Succulent Journal (U.S.)* 61: 104–112.

_____. 1989b. The systematics and evolution of subtribe Steno-cereinae. 8. Organ pipe cactus and its closest relatives. *Cactus and Succulent Journal (U.S.)* 62: 13–24.

_____. 1991a. The Peruvian *Browningia* of *Gymnanthocereus. Cactus and Succulent Journal (U.S.)* 64: 62–68.

_____. 1991b. The systematics and evolution of subtribe Steno-cereinae. 10. The species *Stenocereus griseus. Cactus and Succulent Journal (U.S.)* 62: 161–167.

_____. 1991c. The systematics and evolution of subtribe Stenocereinae. 11. *Stenocereus dumortieri* versus *Isolato-cereus dumortieri. Cactus and Succulent Journal (U.S.)* 63: 92–99.

Gibson, Arthur C., and Karl Horak. 1978. Systematic anatomy and phylogeny of Mexican columnar cacti. *Annals of the Missouri Botanical Gardens* 65: 999–1057.

Gibson, Arthur C., and Park Nobel. 1986. *The cactus primer.* Cambridge, Mass.: Harvard University Press.

Gibson, Arthur C., Kevin Spencer, Renu Bajaj, and Jerry McLaugh-lin. 1986. The ever-changing landscape of cactus systematics. *Annals of the Missouri Botanical Garden* 73: 532–555.

Greenwood, E. 1964. Notas sobre la orientación fototrópica del pseudocefalio en dos especies mexicanos de *Cephalocereus. Cactáceas y Suculentas Mexicanas* 9: 3–6.

Griffen, William B. 1959. *Notes on Seri Indian cultures, Sonora, Mexico.* Latin American Monographs no. 10. Gainesville: University of Florida Press.

Hales, J., and G. Rawley. 1985. The tallest saguaro—hail and farewell. *Cactus and Succulent Journal (U.S.)* 52: 239.

Haselton, Scott E. 1942. King of all cacti. *Cactus and Succulent Journal, Amateur Bulletin Sector* 1: 23–30.

Hastings, J. Rodney, and Robert Humphrey. 1969. *Climatological data and statistics for Sonora and northern Sinaloa.* University of Arizona Institute of Atmospheric Physics Techni-cal Reports on the Meteorology and Climatology of Arid Regions no. 19. Tucson: University of Arizona.

Hastings, J. Rodney, and Ray M. Turner. 1965. *The changing mile.* Tucson: University of Arizona Press.

Herrera, T., and C. Calderón Villagómez. 1994. Cactáceas y agaváceas utilizadas en México para la elaboración de bebidas fermentadas tradicionales. *Cactáceas y Suculentas Mexicanas* 29: 51–58.

Heywood, Vernon. 1993. *Flowering plants of the world.* New York: Oxford University Press.

Hodgson, Wendy. 2001. *Edible plants of the Sonoran Desert.* Tuc-son: University of Arizona Press.

Hunt, David R. 1991. *Pachycereus. Bradleya* 9: 89.

_____. 1992. *CITES Cactaceae checklist.* Kew, England: Royal Botanic Gardens.

Hunt, David R., with Nigel Taylor and Graham Charles. 2006. *The new cactus lexicon.* 2 vols. Milborne Port, England: DH Press.

Joralemon, Donald, and D. Sharon. 1993. *Sorcery and shaman-ism: Curanderos and clients in northern Peru.* Salt Lake City: University of Utah Press.

Kerr, A. C., J. Tarney, G. F. Marriner, A. Nivia, A. D. Saunders, and G. Klaver. 1996. The geochemistry and tectonic setting of late Cretaceous Caribbean and Colombian volcanism. *Journal of South American Earth Sciences* 9: 111–120.

Lamb, Brian. 1991. *Guide to cacti of the world.* New York: Harper Collins.

Lau, Alfred B. 1986. New Ceroid cacti where you do not expect them. *British Cactus and Succulent Journal* 4: 100–102.

_____. 1994. *Neobuxbaumia sanchezmejoradae* Lau. *Cactáceas y Suculentas Mexicanas* 39: 3–7.

Lavender, T. 1989. Yellow-flowered senita. *British Cactus and Succulent Journal* 7: 12.

León de la Luz, José, and Ira Fogel. 2005. *Lophocereus schottii* var. *schottii* forma *spiralis* (Cactaceae) and notes on the monstrose forms. *Cactus and Succulent Journal (U.S.)* 77: 187–189.

Leuenberger, Beat. 1986. *Pereskia* (Cactaceae). *Memoirs of the New York Botanical Garden* 41: 1–141.

Lowe, Charles H., and Warren Steenbergh. 1981. On the Ceno-zoic ecology and evolution of the saguaro. *Desert Plants* 3(2): 82–86.

Mabberley, D. J. 1997. *The plant book.* Cambridge: Cambridge University Press.

MacNeish, Richard. 1967. A summary of the subsistence. In D. Byers, editor, *Prehistory of the Tehuacán Valley,* vol. 1: *Environment and subsistence,* 290–309. Austin: University of Texas Press.

Madsen, Jens. 1989. *Cactaceae.* Vol. 45 of *Flora of Ecuador,* edited by G. Harling and L. Andersson. Copenhagen: Nordic Journal of Botany.

Martin, Paul S., David Yetman, Mark Fishbein, Phil Jenkins, Tom Van Devender, and Rebecca Wilson. 1998. *Gentry's Río Mayo Plants: The tropical deciduous forest and environs of northwest Mexico.* Tucson: University of Arizona Press.

Martínez-Alvarado, D., and A. Flores-Castorena. 1997. Diversidad biológica de la familia Cactáceae en el estado de Morelos, México. *Cactáceas y Suculentas Mexicanas* 42: 7–15.

Mauseth, James D. 1989. Continental drift, climate, and the evolution of cacti. *Cactus and Succulent Journal (U.S.)* 62: 302–308.

Mauseth, James D., Roberto Kiesling, and Carlos Ostolaza. 2002. *A cactus odyssey.* Portland, Ore.: Timber Press.

Mauseth, James D., Teresa Terrazas, Monserrat Vázquez-Sánchez, and Salvador Arias. 2005. Field observations on *Backebergia* and other cacti from Balsas basin, Mexico. *Cactus and Succulent Journal (U.S.)* 77: 132–143.

Méndez, E. 2000. Hibridación natural entre *Trichocereus candicans* y *T. strigosus* en la provincia de Mendoza (Argentina). *Hickenia* 3: 73–76.

Mercado B. A., and S. Granados D. 1999. *La pitaya: Biología, ecología, fisiología sistemática, etnobotánica.* Mexico City: Universidad Autónoma Chapingo.

Meyrán, Jorge. 1970. Las cactáceas columnares de Tehuacán. *Cactáceas y Suculentas Mexicanas* 4: 6–16.

Mitich, Larry. 1997. James Ohio Pattie and the saguaro. *Cactus and Succulent Journal (U.S.)* 69: 33.

Mizrahi, Yosef, Avinoam Nerd, and Park S. Nobel. 1997. Cacti as crops. *Horticultural Reviews* 18: 291–320.

Moran, Reid. 1997. Cardón—personal reminiscence of *Pachycereus pringlei. Cactus and Succulent Journal (U.S.)* 70: 135–147.

Morton, J. F. 1967. Cadushi (*Cereus repandus* Mill.), a useful cactus of Curaçao. *Economic Botany* 21: 185–192.

Mourelle, C., and E. Ezcurra. 1997. Rapoport's Rule: A comparative analysis between South and North American columnar cacti. *The American Naturalist* 150: 131–142.

Murowski, D., T. Fleming, K. Ritland, and J. L. Hamrick. 1994. Mating system of *Pachycereus pringlei*: An autotetraploid cactus. *Heredity* 72: 86–94.

Nabhan, Gary P. 2000. Cultural dispersal of plants and reptiles to the Midriff Islands of the Sea of Cortés: Integrating indigenous human dispersal agents into island biogeography. *Journal of the Southwest* 42: 546–558.

Nerd, Avinoam, E. Ravel, and Yosef Mizrahi. 1993. Adaptation of five columnar species to various conditions in the Negev Desert of Israel. *Journal of Economic Botany* 47: 304–311.

Nerd, Avinoam, Noemi Tel-Zur, and Yosef Mizrahi. 2002. Fruits of vine and columnar cacti. In Park Nobel, editor, *Cacti: Biology and uses,* 185–198. Berkeley: University of California Press.

Nobel, Park. 2002. Physiological ecology of columnar cacti. In Ted Fleming and Alonso Valiente-Banuet, editors, *Columnar cacti and their mutualists: Evolution, ecology, and conservation,* 189–204. Tucson: University of Arizona Press.

Nobel, Park S., and Edward Bobich. 2002. Environmental biology. In Park Nobel, editor, *Cacti: Biology and uses,* 57–74. Berkeley: University of California Press.

Ostolaza, Carlos. 1984. *Trichocereus pachanoi* Br. & R. *Cactus and Succulent Journal (U.S.)* 56: 102–104.

———. 1986. *Browningia candelaris* (Meyen) Br. & R: A new habitat for an old cactus species. *Cactus and Succulent Journal (U.S.)* 58: 13–15.

———. 1997. Etnobotánica IV: La cultural Naxca. *Quepo* 11: 79–86.

———. 2001. El uso del "San Pedro" (*Echinopsis pachanoi*) en medicina tradicional peruana. *Quepo* 15: 28–37.

Ostolaza, Carlos, Larry Mitich, and John King. 1985. *Neoraimondia arequipensis* var. *roseiflora* (Werd. & Backeb.) Rauh. *Cactus and Succulent Journal (U.S.)* 57: 60–64.

Pauley, G. 1984. How big is a saguaro? *Cactus and Succulent Journal (U.S.)* 56: 3.

Pérez de Ribas, Andrés. 1999. *History of the triumphs of our holy faith amongst the most barbarous and fierce peoples of the New World.* Translated from the 1645 manuscript by D. Reff, M. Ahern, and R. Danford. Tucson: University of Arizona Press.

Pfefferkorn, Ignaz. 1989. *Sonora: A description of the province.* Translated by T. Treutlein. Tucson: University of Arizona Press.

Piacenza, Luigi, and Carlos Ostolaza. 2002. Cahuachi y la cultura Nasca. *Quepo* 16: 22–27.

Pimienta-Barrios, Eulogio. 1999. *El pitayo en Jalisco y especies afines en México.* Guadalajara, Mexico: Universidad de Guadalajara.

Pimienta-Barrios, Eulogio, and Park S. Nobel. 1994. Pitaya (*Stenocereus* spp., Cactaceae): An ancient and modern fruit crop of Mexico. *Economic Botany* 48: 76–83.

Pollan, Michael. 2001. *The Botany of Desire.* New York: Random House.

Quimi, Roberto Lindao, and Karen E. Stothert. 1994. *El uso vernáculo de los árboles y plantas en la Península de Santa Elena.* Guayaquil, Ecuador: Fundación Pro-Pueblo.

Reyes Santiago, Jerónimo, Christian Brachet I., Joel Pérez Crisanto, and Araceli Gutiérrez de la Rosa. 2004. *Cactáceas y otras plantas nativas de la Cañada Cuicatlán, Oaxaca.* Mexico City: Sociedad Mexicana de Cactología.

Rudgley, Richard. 1998. *The encyclopedia of psychoactive substances.* London: Little, Brown. Available at: http://www.mescaline.com/sanpedro/

Sahley, C. 2001. Vertebrate pollination, fruit production, and pollen dispersal of *Stenocereus thurberi* (Cactaceae). *The Southwest Naturalist* 46: 261–271.

Salak, M. 2000. In search of the tallest cactus. *Cactus and Succulent Journal (U.S.)* 72: 162–167.

Sánchez-Mejorada R., Hernán. 1970. Viaje a la costa de Jalisco y Colima. *Cactáceas y Suculentas Mexicanas* 15: 29–41.

———. 1972. *Stenocereus chrysocarpus,* una nueva especie de Michoacán. *Cactáceas y Suculentas Mexicanas* 17: 95–98.

———. 1973. The correct name of Grenadier's Cap. *Cactus and Succulent Journal (U.S.)* 45: 171–174.

Scheinvar, Léia, and Hernán Sánchez-Mejorada. 1990. *Neobuxbaumia squamulosa* Scheinvar et Sánchez-Mejorada sp. nova. *Cactáceas y Suculentas Mexicanas* 35: 13–18.

Sharon, D. 1972. The San Pedro Cactus in Peruvian folk healing. In Peter T. Furst, editor, *Flesh of the gods,* 114–135. New York: Praeger.

Shreve, Forrest. 1935. The longevity of cacti. *Cactus and Succulent Journal (U.S.)* 7: 66–68.

Shreve, Forrest, and Ira Wiggins. 1964. *Vegetation and flora of the Sonoran Desert.* 2 vols. Stanford, Calif.: Stanford University Press.

Sobarzo, Horacio. 1961. *Vocabulario sonorense.* Hermosillo, Mexico: Gobierno del Estado.

Soriano, Pascual, and Adriana Ruiz. 2002. The role of bats and birds in the reproduction of columnar cacti in the northern Andes. In Ted Fleming and Alonso Valiente-Banuet, editors, *Columnar cacti and their mutualists: Evolution, ecology, and conservation,* 241–263. Tucson: University of Arizona Press.

Steenbergh, Warren. 1972. Lightning-caused destruction in a desert plant community. *The Southwest Naturalist* 16: 419–429.

Steenbergh, Warren, and Charles Lowe. 1976. *Ecology of the saguaro. I. The role of freezing weather in a warm-desert plant population.* Research in the Parks. Transactions of the National Park Centennial Symposium. National Park Service Symposium Series no. 1. Washington, D.C.: U.S. Department of the Interior.

———. 1977. *Ecology of the saguaro. II.* National Park Service Monograph Series no. 8. Washington, D.C.: U.S. Department of the Interior.

Steward, H. H. 1992. Late Proterozoic and Paleozoic southern margin of North America in northern Mexico. In K. F. Clark, J. Roldán, and R. H. Schmidt, editors, *Geology and mineral resources of the northern Sierra Madre Occidental, Mexico,* 292–299. El Paso: El Paso Geological Society.

Terrazas Salgado, Teresa, and James Mauseth. 2002. Shoot anatomy and morphology. In Park Nobel, editor, *Cacti: Biology and uses,* 23–40. Berkeley: University of California Press.

Turner, Ray, Janice Emily Bowers, and Tony L. Burgess. 1995. *Sonoran Desert plants: An ecological atlas.* Tucson: University of Arizona Press.

Underhill, Ruth. 1946. *Papago Indian religion.* New York: Columbia University Press.

Valiente-Banuet, A., M. D. Arizmendi, A. Rojas Martínez, and L. Domínguez Canseco. 1996. Ecological relationships between columnar cacti and nectar-feeding bats in Mexico. *Journal of Tropical Ecology* 12: 103–119.

Valiente-Banuet, A., M. del Coro Arizmendi, A. Rojas-Martínez, A. Casas, H. Godínez-Álvarez, C. Silva, and P. Dávila-Aranda. 2002. Biotic interactions and population dynamics of columnar cacti. In Ted Fleming and Alonso Valiente-Banuet, editors, *Columnar cacti and their mutualists: Evolution, ecology, and conservation,* 225–241. Tucson: University of Arizona Press.

Valiente-Banuet, Alfonso, and Exequiel Ezcurra. 1991. Shade as a cause of the association between the cactus *Neobuxbaumia tetetzo* and the nurse-plant *Mimosa luisana* in the Tehuacán Valley, Mexico. *Journal of Ecology* 79: 961–971.

Valiente-Banuet, Alfonso, and Hector Godínez-Álvarez. 2002. Population and community ecology. In Park Nobel, editor, *Cacti: Biology and uses,* 91–108. Berkeley: University of California Press.

Valiente-Banuet, A., G. F. Vite, and A. Zavala-Hurtado. 1991. Interaction between the cactus *Neobuxbaumia tetetzo* and the nurse shrub *Mimosa luisana. Journal of Vegetation Science* 2: 12–14.

Van Devender, Thomas R. 1990. Late Quaternary vegetation and climate of the Sonoran Desert, United States and Mexico. In J. L. Betancourt, T. R. Van Devender, and P. S. Martin, editors, *Packrat middens: The last 40,000 years of biotic change,* 134–165. Tucson: University of Arizona Press.

———. 2002. Environmental history of the Sonoran Desert. In Ted Fleming and Alfonso Valiente-Banuet, editors, *Columnar cacti and their mutualists: Evolution, ecology, and conservation,* 3–24. Tucson: University of Arizona Press.

Vega-Villasante, F., H. Áviles, G. Montaño, I Gómez, J. L. Espinoza, K. Busto, H. Romero-Schmidt, and H. Nolasco. 1995. Mexican cactus ethnobotany: The contribution of cacti to the survival of the natives of Baja California. *Cactus and Succulent Journal (U.S.)* 67: 74–79.

Wallace, Robert. 2002. The phylogeny and systematics of columnar cacti: An overview. In Ted Fleming and Alonso Valiente-Banuet, editors, *Columnar cacti and their mutualists: Evolution, ecology, and conservation,* 42–65. Tucson: University of Arizona Press.

Wallace, Robert, and Arthur C. Gibson. 2002. Evolution and systematics. In Park Nobel, editor, *Cacti: Biology and uses,* 1–22. Berkeley: University of California Press.

Weiner, Jonathan. 1994. *The beak of the finch.* New York: Knopf.

Yetman, David. 1996. *Sonora: An intimate geography.* Albuquerque: University of New Mexico Press.

———. 2002. *The Guarijíos of the Sierra Madre: Hidden people of northwestern Mexico.* Albuquerque: University of New Mexico Press.

———. 2004. Columnar cacti of the Río Marañon, Peru. *Cactus and Succulent Journal* 76: 15–27.

Yetman, David, and Alberto Búrquez M. 1996. A tale of two species: Speculation on the introduction of *Pachycereus pringlei* in the Sierra Libre, Sonora, Mexico, by *Homo sapiens. Desert Plants* 12: 23–32.

Yetman, David, and T. R. Van Devender. 2002. *Mayo ethnobotany: Land, history, and tradition in Northwest Mexico.* Berkeley: University of California Press.

Zappi, Daniella. 1994. Pilosocereus *(Cactaceae): The genus in Brazil.* Succulent Plant Research no. 3. Sherborne, England: David Hunt.

General Index

SCIENTIFIC NAMES INDEX

THE SOUTHWEST CENTER SERIES

Joseph C. Wilder, editor

ABOUT THE AUTHOR

David Yetman is a research social scientist at the University of Arizona's Southwest Center. He has been traveling in Latin America since 1961 and has written extensively about native peoples and plants of northwestern Mexico. His publications include *Sonora: An Intimate Geography* (1996) and *Guarijíos of the Sierra Madre: Hidden Peoples of Northwest Mexico* (2002). Yetman confesses to a decades-old fascination with columnar cacti that led to the publication of this book. Among his long-term projects is a joint effort with the Masiaca Indigenous Community (Mayo) of southern Sonora, Mexico, to produce marketable food products from organ pipe cacti, which abound in the region. He is also host of the Public Broadcasting System series *The Desert Speaks*. A self-described "desert rat," Yetman lives in Tucson, Arizona, his yard densely planted with tall cacti.